CLASSI

The north of Ireland

CLASSIC GEOLOGY IN EUROPE SERIES

1. *Italian volcanoes* Chris Kilburn & Bill McGuire
2. *Auvergne* Peter Cattermole
3. *Iceland* Thor Thordarson & Armann Hoskuldsson
4. *Canary Islands* Juan Carlos Carracedo & Simon Day
5. *The north of Ireland* Paul Lyle
6. *Leinster* Chris Stillman & George Sevastopulo
7. *Cyprus* Stephen Edwards, Karen Hudson-Edwards, Joe Cann, John Malpas, Costas Xenophontos
8. *The Dalradian of Scotland* Jack Treagus
9. *The Northwest Highlands of Scotland* Con Gillen
10. *The Inner Hebrides of Scotland* Con Gillen

OTHER TITLES OF RELATED INTEREST

Confronting catastrophe: new perspectives on natural disasters
David Alexander

Geology and landscapes of Scotland
Con Gillen

La catastrophe: Mount Pelée and the destruction of Saint-Pierre, Martinique
Alwyn Scarth

Monitoring the Earth: physical geology in action
Claudio Vita-Finzi

Principles of emergency planning and management
David Alexander

Volcanoes of Europe
Alwyn Scarth & Jean-Claude Tanguy

Wind and landscape: an introduction to aeolian geomorphology
Ian Livingstone & Andrew Warren

CLASSIC GEOLOGY IN EUROPE 5

The north of Ireland

Paul Lyle
University of Ulster

TERRA

First published in 2003 by Terra Publishing

Terra Publishing
PO Box 315, Harpenden, Hertfordshire AL5 2ZD, England
Telephone: +44 (0)1582 762413
Fax: +44 (0)870 055 8105
Website: http://www.terrapublishing.co.uk
E-mail: publishing@rjpc.demon.co.uk

ISBN: 1-903544-08-4 paperback

12 11 10 09 08 07 06 05 04 03
10 9 8 7 6 5 4 3 2 1

British Library Cataloguing-in-Publication Data
A CIP record for this book is available from the British Library

Library of Congress Cataloging-in-Publication Data are available

This publication is supported by the Environment and Heritage Service, Northern Ireland

Typeset in Palatino and Helvetica
Printed and bound in Italy by EuroGrafica SpA.

Contents

Preface

The north of Ireland is renowned for its unspoiled scenery and has also long been recognized for the diversity and ease of access of its geology. The publication by the Geological Survey of Northern Ireland in 1997 of a revised edition of the 1:250 000 scale map of the geology of Northern Ireland was seen as an opportunity to produce a geological guide book to the north of Ireland, based on this map. Included are some of the localities that have featured in the great geological controversies of the past, such as the Giant's Causeway and the Donegal granites. Increasing numbers of visitors are coming to the area each year, many of them seeking to explore the geological foundation of the landscape. This book aims to explain and illustrate that foundation and, although written primarily for the non-specialist reader, I hope the treatment is sufficiently rigorous for use by professional geologists and student groups coming to one of the classic areas for geological fieldwork in western Europe. For those with little previous knowledge of geology a glossary of geological terms is included.

The first part of the book covers the geological history of Ireland within the context of the formation of the Earth and plate-tectonic processes and deals with geological time, the interpretation of geological maps and the formation and composition of the principal rock types. The north of Ireland has a wealth of archaeological remains, so the relationship between geology and archaeology is explored, in particular the role of rocks in the social, cultural and industrial development of the inhabitants of the area, from the earliest settlers to the present day. The second part of the book contains ten detailed field-trip itineraries, illustrated with maps, diagrams and field photographs. I trust that this will allow all readers, irrespective of their geological background, to examine the evidence and draw conclusions at first hand, thereby enhancing their appreciation of the landscape. The excursions have been chosen to illustrate the wide range of geological features across the whole region, with examples from Donegal and Sligo, Fermanagh and Tyrone, north and east Antrim and north and south Down.

Paul Lyle
Newtownards
March 2003

Acknowledgements

I am pleased to acknowledge those people without whose help this book could not have been written. Much of the work was done during a sabbatical year spent as Academic Visitor in the Department of Geology at Trinity College, Dublin, and I would like to thank all the staff and graduate students there for their hospitality and encouragement, particularly George Sevastopulo and Chris Stillman. I am grateful to my colleagues at Jordanstown for their support and their tolerance of my absence during that same period.

The geologists at the Geological Survey of Northern Ireland (GSNI) have been extremely helpful, particularly Derek Reay and Terry Johnston, and I have drawn extensively on GSNI maps and publications for the excursion itineraries. For those itineraries in the Republic of Ireland I gratefully acknowledge the support and cooperation of the Geological Survey of Ireland (GSI). Barry Long and Brian McConnell of GSI, John Arthurs of GSNI, and Mike Simms of the Ulster Museum all read sections of the text and provided much useful and constructive comment. The Environment and Heritage Service of the Department of the Environment is thanked for its financial and other support – which has made it possible to print the book in colour – and I am particularly indebted to Ian Enlander for his efforts on behalf of the whole project.

My thanks go to all those colleagues and friends for their company in the field and their freely given opinions and advice on the geology of the north of Ireland and the excursion itineraries. Notwithstanding such sterling work however, any errors and omissions are my responsibility entirely.

I am grateful to Kilian McDaid, who created the diagrams, and to Jennifer Larkin, who deciphered and typed the manuscript. Sylvia Lyle read several versions of the manuscript in detail and visited all of the field localities with me. As always, her perceptive and infuriating comments helped to clarify the text throughout the book, for which I am eternally grateful. Finally, I would like to dedicate this book to Gareth, Rachel and Simon for their heroic patience and stoicism displayed over the years while waiting for me to finish looking at outcrops of rock.

ACKNOWLEDGEMENTS

Topographical maps are reproduced by permission of the Ordnance Survey of Northern Ireland on behalf of the Controller of Her Majesty's Stationery Office; © Crown copyright 2003, permit 20209. Extracts from the 1:250 000 *Geological map of Northern Ireland (Solid Edition)* are reproduced by permission of the Geological Survey of Northern Ireland; © Crown copyright, GSNI permit 2003/1. Maps including parts of the Republic of Ireland are based on Ordnance Survey Ireland permit MP005902; © Ordnance Survey of Ireland and Government of Ireland.

Chapter 1

Introduction

The northern third of Ireland contains a remarkable diversity of rock types covering an age range from 1500 million years ago to the present day, from Precambrian to Recent. This range of rocks within an area some 200×150 km, offers an extensive array of geological environments to be examined with relative ease. Many of the localities, such as the Giant's Causeway and the Donegal granites, featured in some of the great geological controversies of the past and are important in the development of geology as a science since the eighteenth century.

This book aims to describe the geological history of the north of Ireland in terms of the theory of **plate tectonics**, which provides a global framework for geological processes, and to demonstrate this geological diversity by illustrated geological excursion guides to the most important localities in the region.

The attractive scenery is what draws most of the visitors to Ireland and what residents most appreciate about living here permanently. An appreciation of the fundamentals of the landscape, how it came to be formed and what processes have sculpted it allows an interpretation of the scenery at a deeper level of understanding. Percy French described the Mountains of Mourne as "sweeping down to the sea" in his famous song. The description is a very apt one, but why do the granite mountains of the Mournes have that very distinctive rounded profile, which does indeed seem to sweep towards the sea at Newcastle, and why are they there in the first place? To answer these questions satisfactorily requires an understanding of geological processes acting in south Down from around 400 million years ago, which produced sandstones and mudstones on the bed of an ocean that has long disappeared. After hundreds of millions of years, while these rocks were involved in a complex mountain-building episode, during which they were compressed and fractured, continental splitting produced a large volume of molten rock known as **magma**. This forced its way up into the sandstones and mudstones, to form the granites of the Mourne Mountains. Although these granites solidified deep in the crust at about 60 million years ago, erosion since that time has removed the covering of softer sandstones and mudstones and left the rounded shapes of the harder granites standing proud from the softer sedimentary rocks. Granite mountains typically have a rounded profile because the mineral feldspar, which is an important constituent of granite, weathers out readily. Every element of the Irish landscape can be explained in similar terms, with processes acting globally and over very long periods of time.

Ireland is not static on the Earth's surface. The theory of plate tectonics suggests that the outer layers of the Earth are dynamic and continents are moving or drifting over its surface. The ocean that produced the sandstones and mudstones around the Mourne granites was south of the Equator 400 million years ago, probably somewhere near the longitude of Indonesia, and has since contracted and disappeared. By the time the granites of the Mournes were being intruded, that part of the crust we now recognize as Ireland had drifted east and north around the Earth to somewhere around the position presently occupied by the south of France. In the 60 million years since the Mournes were first formed, Ireland has moved north to its present position, and is likely to go on drifting towards polar latitudes in the future.

How geologists are able to make such statements, the evidence that exists for the various locations of Ireland on the Earth's surface throughout geological time and the timetable of global movements will be discussed in later chapters. The treatment is based on the 1:250 000 geological map of Northern Ireland, including Donegal and parts of Sligo, Leitrim, Cavan, Monaghan and Louth, as well as the six counties of Northern Ireland, and is aimed at those interested in exploring the geology of this part of Ireland, whether as amateurs, students or professionals, as individual visitors or as part of larger groups.

Chapter 2

Ireland from space

Since the development of space travel in the 1960s, we have become increasingly accustomed to images of the Earth viewed from space, taken by satellite, at varying distances above the surface. These have proved extremely useful to meteorologists, enabling them to see the large-scale features of weather systems, thereby helping in their forecasts. However, they have also been utilized by geologists as an invaluable addition to the information collected by traditional fieldwork methods. It is often much easier to interpret the large-scale structural features of the crust when viewed obliquely or vertically from above. As well as conventional cameras, sensors that detect wavelengths outside the visible light spectrum, such as infrared radiation, can often detect patterns invisible to the human eye.

Geographically, Ireland is one of a group of islands (an archipelago) off the northwestern edge of the European continent. In geological terms it is very much part of that European continent, and the continental shelf of Europe, on which Ireland is situated, extends far off shore from Ireland under the waters of the Atlantic Ocean.

A satellite image of the north of Ireland (Fig. 2.1) reveals much of the structure of the island. The area can be subdivided into natural regions, of which the most prominent are the Donegal Highlands, the lavas of the Antrim Plateau, the Mourne Mountains, the Sperrin Highlands, the Fermanagh Lakelands and the valleys of the River Bann and River Lagan. The differences in topography and vegetation across these regions are clearly discernible from space and reflect variations in the composition and structural history of the bedrock in the regions. Later sections of this book aim to explain the global contexts in which many of these rocks were formed and their subsequent geological history.

We know now that the surface of the Earth is not fixed, but continents move across the surface, and oceans open and close, and that Ireland has had a long and complex journey around the world before it arrived at its present position. For example, many of the surface features of the north of Ireland, such as the Donegal Highlands, are aligned northeast–southwest.

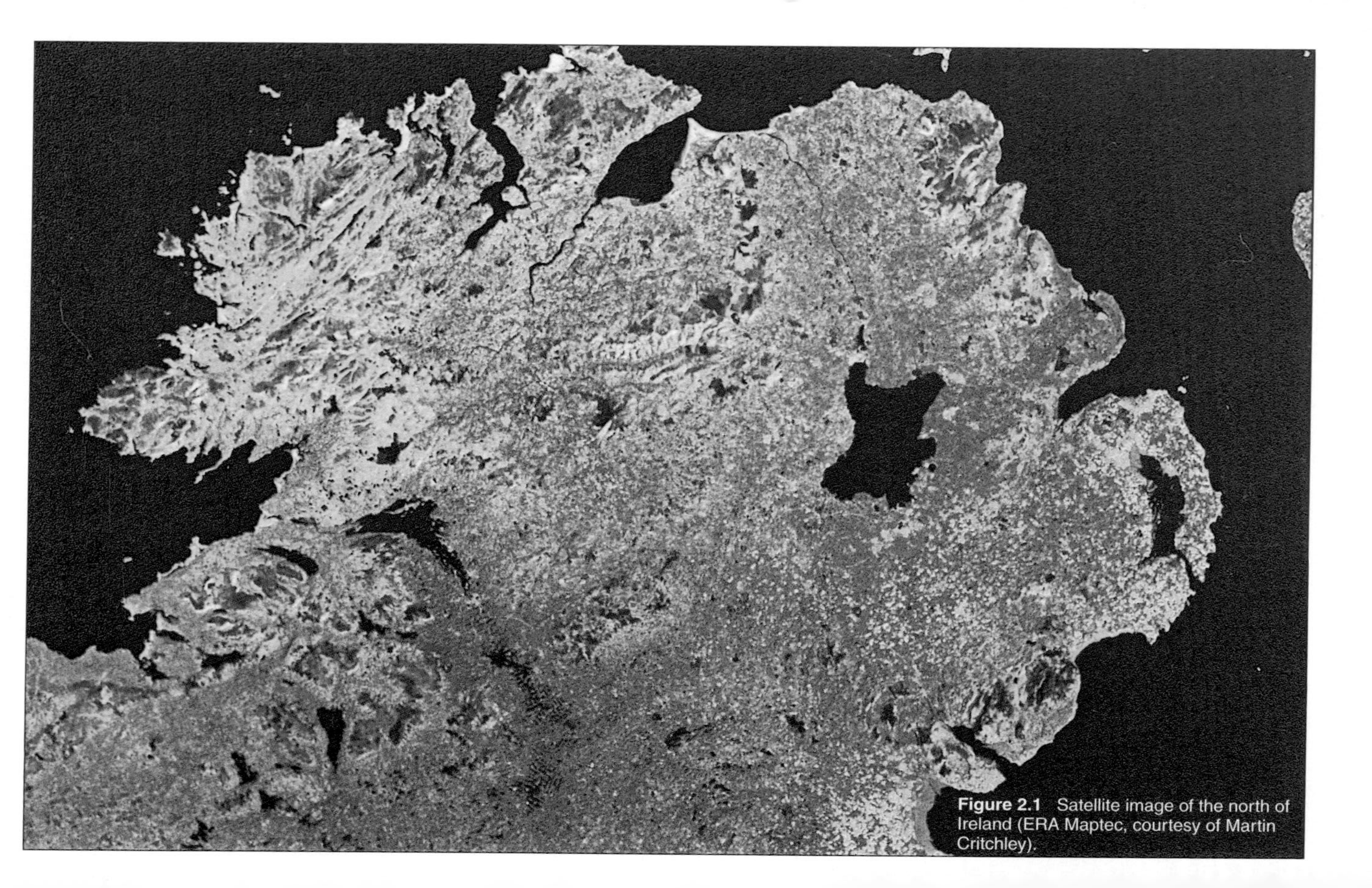

Figure 2.1 Satellite image of the north of Ireland (ERA Maptec, courtesy of Martin Critchley).

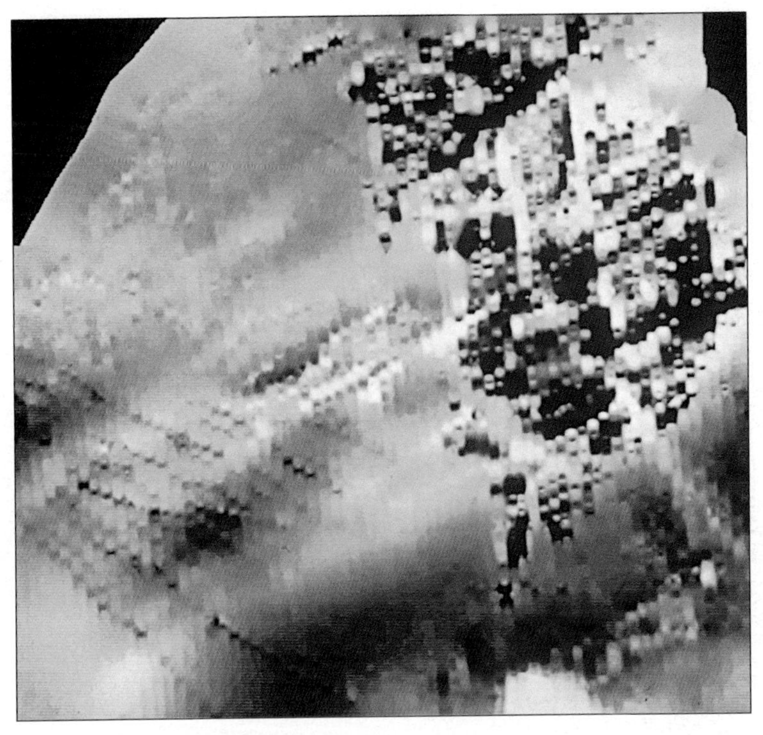

Figure 2.2 Variations of the geomagnetic field over part of the north of Ireland (data courtesy of the Geological Survey of Northern Ireland, processed by P. J. Gibson).

This reflects processes of mountain building some 400 million years ago when Ireland was caught up in the collision of two continents as an ocean closed and a range of high mountains, the Caledonides, was formed. More recently, geologically speaking, the Himalayas have formed in a similar way from the collision of India and Asia.

Other measurements that can be taken over the whole region from a distance, from aircraft or satellites, are variations in the magnetic and gravitational fields of the Earth. When these data are processed it is often possible to recognize important features of the crust. For example, Figure 2.2 shows variations in the geomagnetic field for a section of the north of Ireland covering counties Antrim, Down and parts of Tyrone, Fermanagh and Armagh. The Antrim basalts to the east contain a high percentage of a natural magnetic mineral called **magnetite**, and so they produce a characteristically dense pattern on the image, compared with the textured appearance of much of the rest of the image. In the southwest part of the image there is a series of near-parallel linear features running roughly northwest–southeast. These are rocks of a composition very similar to the

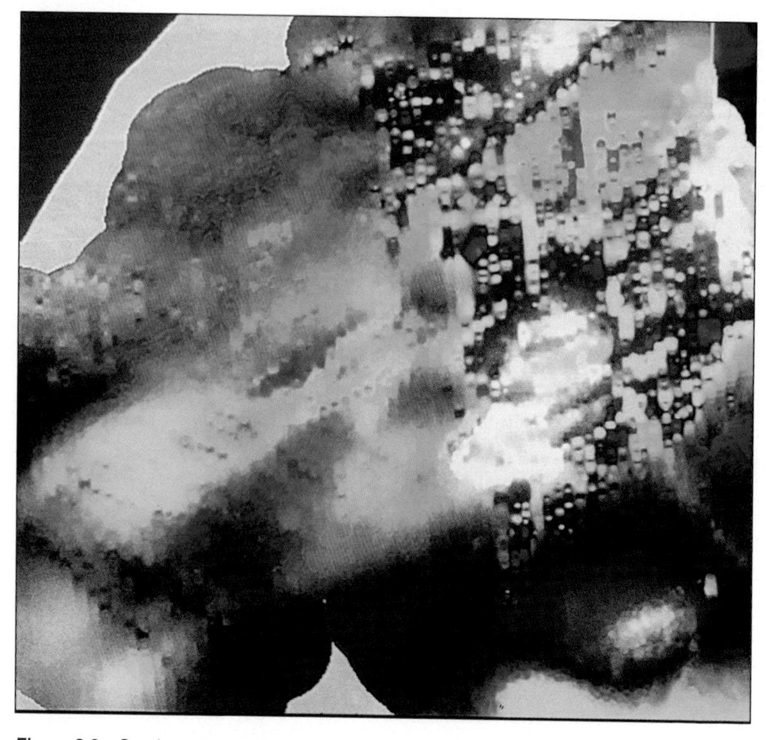

Figure 2.3 Gravity and geomagnetic variation over part of the north of Ireland (data courtesy of the Geological Survey of Northern Ireland, processed by P. J. Gibson).

Antrim basalts, which is why they have a similar appearance on the geomagnetic image. They occur in Fermanagh, and what is interesting about this image is that it indicates a much more extensive occurrence of these rocks in Fermanagh than is evident on the surface. The recognition of subsurface features such as these is an important technique in exploration geology in the continuing search for mineral or hydrocarbon reserves.

More data can be added to this image to provide further information on the substructure of the north of Ireland. The basalt lavas that make up the Antrim Plateau are denser than the granites of the Donegal Highlands or the Mourne Mountains. This means that the Earth's gravitational field tends to be lower in granite areas than in basalt areas, and this variation is reflected in the range of colours shown on Figure 2.3. Here darker shades such as blue indicate higher-density rocks with a high gravity signal, whereas shades of red and yellow indicate the presence of less dense rocks such as sediments and granite. The pinkish-brown area to the north of the image indicates a thick sedimentary basin under the basalts in north Antrim, and the Lough Neagh basin is also marked by a distinct pink area.

The granites of the south Down area show up as a prominent red signature, and the near-parallel linear features noted in Fermanagh on the magnetic image are seen here to be underlain by a deep blue shade. This suggests that, at depth, a body of basaltic rock exists that has been the source of the basaltic rocks near the surface. Using a composite image such as Figure 2.3, which has been produced by combining the two datasets and displaying them simultaneously, it is possible to add to the interpretation of the geology based on more conventional methods.

Chapter 3

Earth formation and plate tectonics

The Earth is a planet in the Solar System, in orbit around a star. This star, which we call the Sun, is one star out of millions in our galaxy, which is one of countless millions of galaxies in the Universe. Thus, the Earth is a very small component in the universe, no matter how significant it may appear to those who live on its surface. The Solar System, including the Earth, is probably about 4600 million years old and it consisted originally of a cloud or nebula of gases and dust particles from which the Sun and planets gradually condensed and formed. The result was an Earth consisting of concentric layers, with the densest part, the **core**, in the centre, and the lightest part, the **crust**, on the outside. Between the core and crust is the **mantle**, a thick zone of mostly solid rock, but a solid that is capable of slow movement in the way that ice in a glacier can flow while still remaining solid. The ability of the mantle to move has been important in the geological history and development of the Earth.

Plate-tectonic processes

It is important to realize that the Earth as a system is not static but constantly changing. This applies not only to its outer surface, hydrosphere and atmosphere, which can be readily seen to change, but also to its internal layers such as the deep crust and the mantle. Ireland is currently located about 6° west of the 0° line of longitude (the Greenwich meridian) and around 55°N latitude. This has not always been the case. It is now known that the surface of the Earth is dynamic and composed of polygonal segments called **plates**. Plate tectonics is the name given to the process whereby these large segments move slowly in relation to each other, at around a few centimetres per year. Plates move towards or away from each other, or slide past each other, and thus, on the surface of the Earth, oceans are slowly expanding or contracting, and continents are moving or

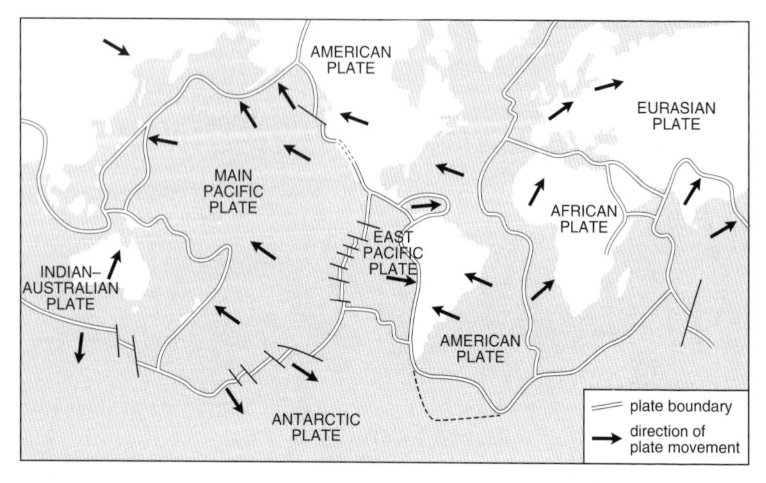

Figure 3.1 The distribution and direction of movement of the principal tectonic plates on the Earth's surface.

drifting. The junctions between adjacent plates are sites of earthquakes and volcanoes, and the nature and frequency of earthquakes and volcanic eruptions depend on the directions of movement of the plates (Fig. 3.1).

The general directions and rates of plate movements are now quite well understood by Earth scientists; however, there is still some debate over the precise causes of plate motion. As mentioned earlier, the Earth can be subdivided into concentric layers. The outer relatively thin and brittle layer is the crust, which overlies a thick layer going down to almost 3000 km from the surface, the mantle, which although solid is hot and capable of flowing very slowly like a viscous liquid. The central zone is the core, consisting predominantly of iron, with a liquid outer part and a solid inner part (Fig. 3.2). The moving plates are formed of the continental or oceanic crust, along with the brittle upper part of the mantle, and these regions together are known as the **lithosphere**, or rocky layer. The region directly below the lithosphere is the **asthenosphere**, or weak layer, and is partially liquid because in places the mantle has begun to melt. It is probably this relatively weak layer along which the lithospheric plates can move or "drift". The precise causes of plate movement are likely to involve several factors, but it is likely that the process of **mantle convection** is at least part of the driving mechanism. Convection is the movement of material as a result of temperature differences. For example, a saucepan of soup being heated on a hob will have convection currents set up within it. Heat from the burner enters the soup at the bottom of the pan and is carried upwards by the warmed soup, which rises because it is less dense, or lighter, than the colder unheated soup. However, this hotter material cools at the surface

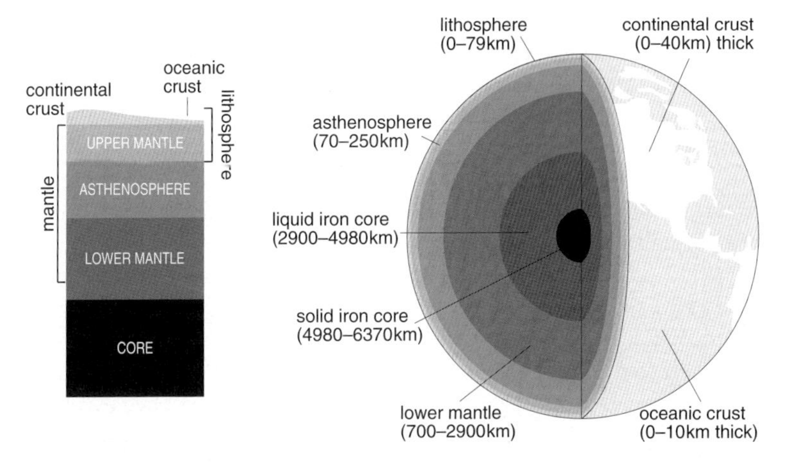

Figure 3.2 The main internal divisions of the Earth.

and thus becomes heavier, and then sinks to the bottom of the pan, to be recycled as part of a **convection cell** when it returns to the surface of the pan after re-heating (Fig. 3.3a). The same process is thought to operate within the mantle, where the heat is from radioactive minerals deep within the Earth's interior (Fig. 3.3b). In simple terms, the rising parts of convection cells occur underneath those parts of the crust where the plates are moving apart; since the crust is being stretched and new crust is being formed, these regions are referred to as **constructive plate margins** (Fig. 3.3c). Many of these constructive plate margins are now in the oceans and

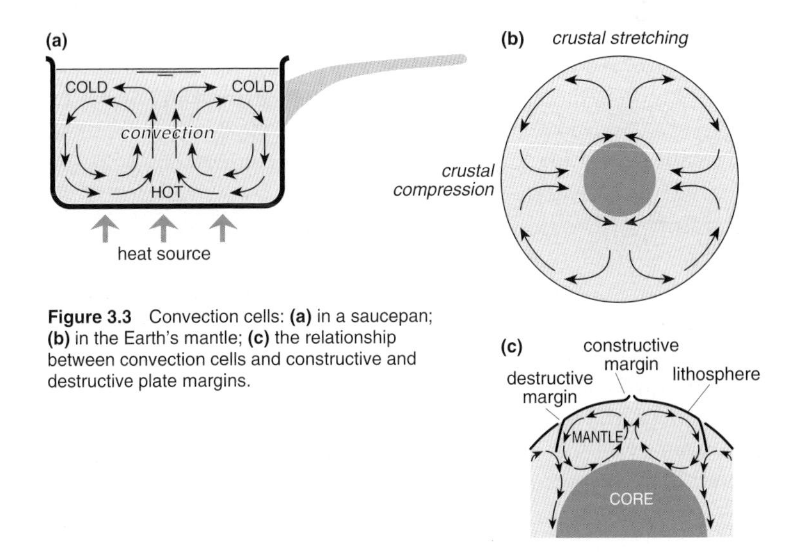

Figure 3.3 Convection cells: **(a)** in a saucepan; **(b)** in the Earth's mantle; **(c)** the relationship between convection cells and constructive and destructive plate margins.

occur as vast submarine mountain ranges known as the **mid-ocean ridges** (Fig. 3.4). The sinking or descending limbs of the convection cells occur where plates are moving together, such as around the Pacific Ocean. Since the crust here is being compressed and shortened by the heavier oceanic plate being forced down below the lighter continental plate (Fig. 3.3c), they are described as **destructive plate margins**.

On a large scale, therefore, it seems that the movement of plates is part of an overall convection system in the mantle. There is upward movement of melted mantle occurring under constructive plate margins, followed by a sideways or lateral movement, and then a descent of the plate over the downward limb of the convection cell. The process whereby the oceanic plate is forced under the continental plate is referred to as **subduction**. This process leads to intense folding and deformation, and eventually to partial melting of the crust to produce a **granite** liquid. The various granite **batholiths** found in Donegal were formed in this environment. The cycle from constructive plate margin at the mid-ocean ridges to destructive plate margin at the subduction zone is summarized in Figure 3.4.

Earthquakes and volcanic eruptions occur at both types of margin. Although the activity at constructive margins tends to be relatively subdued, the volcanic eruptions and the earthquakes associated with destructive plate margins, such as the Andes and Indonesia, are among the most violent and destructive in nature. Volcanoes at constructive plate margins are described as **effusive** and tend to produce fluid magmas of basaltic composition, whereas those at destructive plate margins are **explosive** and produce more viscous lava of **andesite** or **rhyolite** composition. Where two continental plates are in collision (e.g. India colliding with Asia), the crust is compressed and buckled, and both plates are forced upwards to form high mountains such as the Himalayas. Incidentally, the earthquakes associated with the famous San Andreas Fault in the Los Angeles–San Francisco area are associated with the third type, called a **conservative plate margin**: here the plates are neither converging nor diverging but are sliding past each other.

Superimposed on this large-scale mantle circulation is a smaller-scale system of up-currents, often referred to as **mantle plumes**. These mantle plumes or hotspots account for the volcanic activity in localities such as Hawaii, far away from plate margins. They are also important in starting continental break-up during plate-tectonic activity. As mantle plumes reach the surface under a continent, the continental crust swells upwards because of the heat from the plume. This swelling creates stress that can be relieved only by crustal fracturing and rifting. This process resulted in the separation of Europe from North America and Greenland about 60 million years ago, with the eruption at that time of the Antrim basalts.

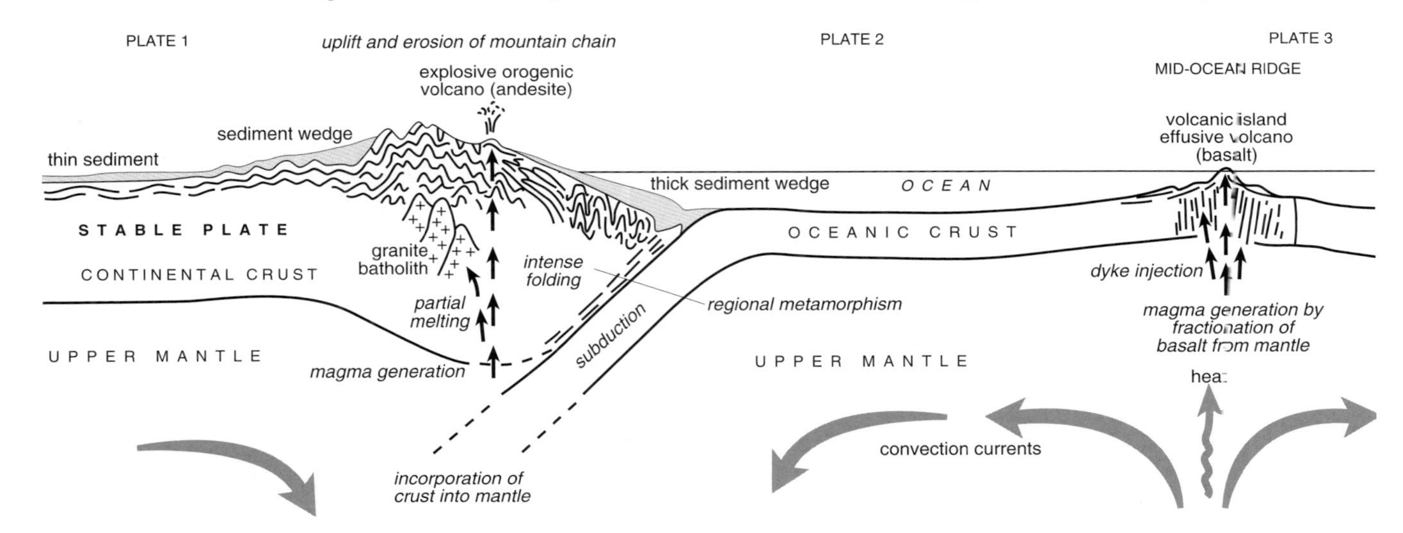

Figure 3.4 Summary of plate tectonic cycle.

Rock types and their formation

Various rock types such as basalt and granite have already been mentioned. It is useful at this stage to look at the formation and classification of those rock types most commonly met, before looking at the geological history of Ireland in greater detail. Although there are literally hundreds of different rocks recognized by geologists, it is possible to understand most geological processes by reference to a limited number of types. There are three main groups: **igneous**, **sedimentary** and **metamorphic**.

Igneous rocks

Igneous rocks (Latin *ignis*: "fire") are formed by the cooling of hot molten material known as magma, from the Earth's interior. This magma, which can have various compositions, forms by the melting of deep crustal or mantle rocks to produce a hot liquid that is less dense than the surrounding solid rock. This difference in density causes the magma to rise towards the surface in the way that oil globules rise to the surface in water. If the magma is prevented from reaching the surface and cools within the crust, the resultant igneous rock is called **intrusive**. It cools slowly and the crystals in the rock are therefore relatively large and visible to the naked eye, since the slower the cooling, the larger the crystals that form. Granite is a good example of an intrusive igneous rock. If the magma reaches the surface, usually in the form of a lava flow erupted from a volcano, it forms an **extrusive** rock and cools quickly on exposure to air temperatures. This means the crystals are much smaller than those of intrusive rocks and are often impossible to distinguish with the naked eye. **Basalt** is a widespread example of extrusive igneous rock. If the magma cools very rapidly, perhaps by coming into contact with water, for instance, then there is little or no time available for crystals to grow and therefore a volcanic glass is formed, of which **obsidian** is a good example. If basaltic magma does not reach the surface and cools instead at depth, it forms **gabbro**, the coarse-grain equivalent of basalt. Intermediate in grain size between gabbro and basalt is **dolerite**, frequently occurring in the relatively small-scale vertical **dykes** and horizontal **sills** that cut through the surrounding rock. Similarly, if the granite magma considered earlier does make it to the surface, it will also form a lava flow. This type of lava is referred to as **rhyolite** and is usually light grey or pink, contrasting with the dark grey or black of basalt. A simple classification scheme for the igneous rocks is shown below in Table 3.1. Classification is based on the colour of the rock (its colour index), varying from pale at the granitic end of the range to dark at the basaltic end. The colour reflects the relative proportions of light and dark minerals, which in turn reflect the overall chemical composition of

Table 3.1 Classification of igneous rocks.

Grain size		Type of rock	
Coarse grain > 2.0 mm	Gabbro	Diorite	Granite
Medium grain 2.0–0.06 mm	Dolerite	Microdiorite	Microgranite
Fine grain 0.06–0.002 mm	Basalt	Andesite	Rhyolite
<0.002 mm	Volcanic glass	Volcanic glass	Obsidian
Colour index	Dark	Medium	Pale
Silica content	Basic Little or no quartz	Intermediate Some quartz	Acid Much quartz

the rocks. For example, the basaltic rocks tend to be rich in elements such as iron and magnesium, which concentrate in dark minerals such as pyroxene, whereas granitic rocks tend to have higher concentrations of the elements silicon, aluminium and potassium, which concentrate in pale minerals such as feldspar and **quartz**.

Landforms of igneous rocks Figure 3.5 shows the commonest landforms produced by intrusive and extrusive igneous activity. Large-scale intrusions are referred to as batholiths or **plutons**, forming deep within the crust and often covering very large areas, for example the various granite plutons found in Donegal. Rocks intruded by bodies of magma are referred to as **country rocks**. Plutons may also act as **magma chambers**, feeding minor intrusions higher in the crust or volcanic eruptions on the surface. These minor intrusions are referred to as dykes if they are vertical and cut across existing sedimentary beds, or sills if they are horizontal and parallel to existing structures. Magma cooling at the surface forms **lava flows**; a series of lava flows can build a volcanic cone or a lava plateau such as the Antrim basalts. Many volcanic cones are **composite**: they are composed of layers of **ash** and lava. The solidified magma left in the feeder pipe of the volcano at the end of an eruption often forms a volcanic neck or **plug** after the rest of the cone has been eroded.

A further contrast between the basaltic and granitic rocks is in viscosity, which is also manifested in landforms. Basalt is fluid and relatively fast moving; rhyolite is very viscous and slow moving. All magmas contain a proportion of dissolved gases; as the magma moves towards the surface, these gases expand as the surrounding pressure is lowered. The high viscosity of rhyolite magma means that it is difficult for expanding volcanic gases to escape from the liquid and, consequently, internal pressure builds up until the magma explodes, often catastrophically, as happened recently at Mount Pinatubo in Indonesia. This explosive nature of rhyolite volcanoes is in contrast to the so-called effusive basalt volcanoes such as

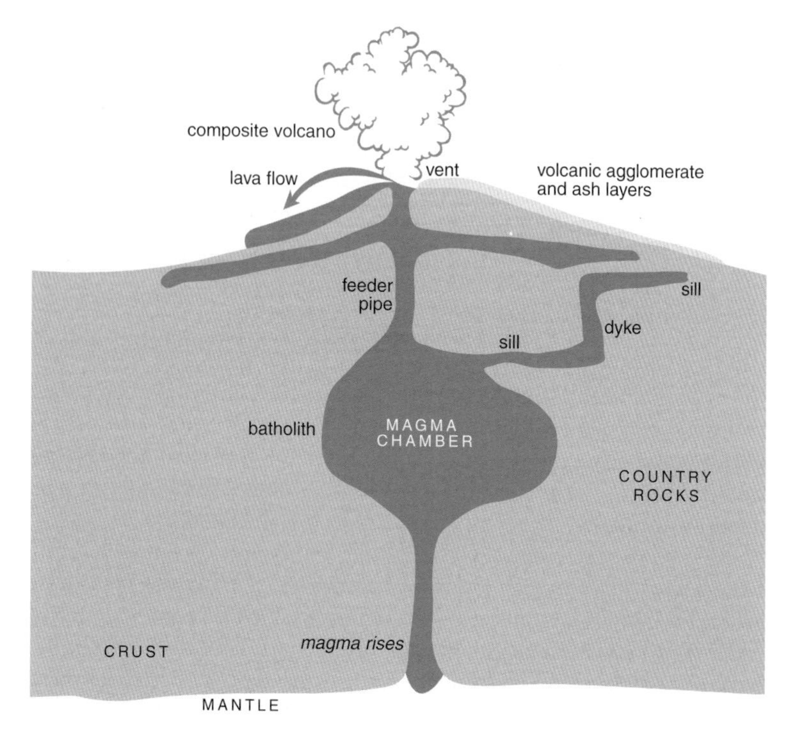

Figure 3.5 Landforms produced by intrusive and extrusive igneous activity.

Hawaii. Here the greater fluidity of the magma allows the gases to escape, often producing spectacular lava fountains, rather like the effusions of champagne when the cork is suddenly released from the bottle.

Sedimentary rocks

When landforms have been produced from igneous rocks, whether intrusive or extrusive, those landscapes become subject to the forces of weathering and erosion. Weathering agents such as rain, groundwater, frost and wind alter the rocks physically and chemically to produce a weakened **regolith**: a covering of loose material. This can then be removed or eroded by ice, wind or water, and it moves steadily down hill under gravity. These are stages in the process of formation of the **clastic sediments**: those made up of broken fragments of pre-existing rocks. Sedimentation is the deposition of such materials in the lowest places to which water and air currents can carry them. Figure 3.6 shows the downhill movement under gravity of the weathered particles, and some of the different environments of deposition en route. Clastic sediments are classified according to their grain size because sedimentation of the eroded particles is mainly controlled

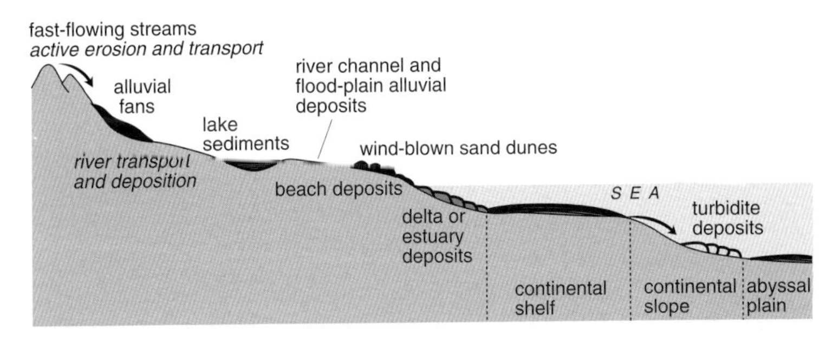

Figure 3.6 The downhill movement and deposition under gravity of weathered particles.

by the strength of the water or air current that carries them (Table 3.2).

The mineralogical components of these rocks are mainly quartz, feldspar and clay. A range of environments of deposition of clastic sediments will be explored in the field excursions.

Figure 3.7 shows the velocity of water current required to carry a range of sedimentary particles from clay to sand and gravel. Faster water currents are capable of moving a wider range of particle sizes and larger grain sizes than area slower currents. For example, a water current of 0.1 m per second can transport particles up to 1 mm in diameter (sand size), whereas a current of 0.01 m per second can transport grains up to only 0.0125 mm in diameter (silt size). When the river gradient decreases and the velocity drops, the current will progressively deposit the material, the coarsest grains being the first to be dumped. This leads to the deposition of sediment in layers as the current velocity fluctuates. The energy of the current is indicated by the grain size of the rocks, from mud and clay at the fine end of the range, through siltstones, sandstones and conglomerates as grain size increases. The oldest sediments will be deposited at the bottom of the succession, which becomes progressively younger upwards.

The other principal sedimentary rocks are the chemical sediments that are precipitated from solution, mostly in the oceans. The most abundant

Table 3.2 Classification of clastic sediments by grain diameter (mm).

Grain diameter	Rock type
Gravel >2.0	Conglomerate
Very coarse sand 1.0–2.0	
Coarse sand 0.5–1.0	
Medium sand 0.25–0.5	Sandstone
Fine sand 0.125–0.25	
Very fine sand 0.0625–0.125	
Silt 0.004–0.0625	Siltstone
Clay <0.004	Claystone

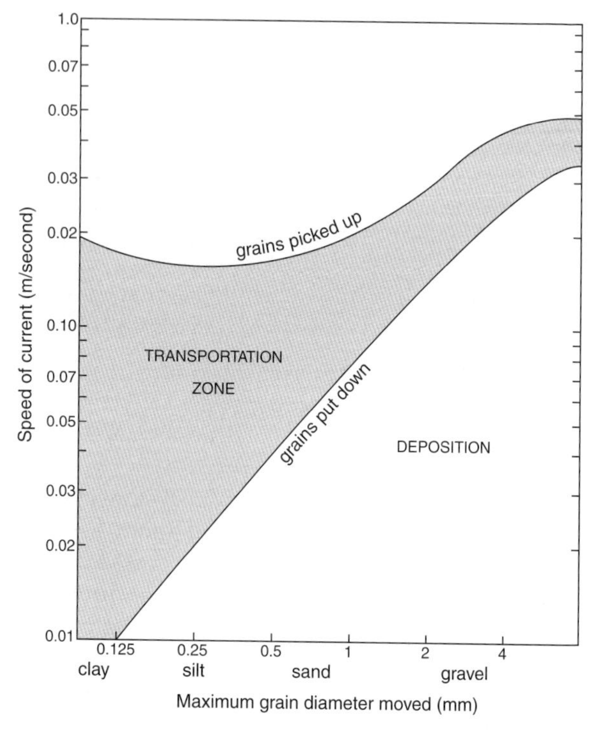

Figure 3.7 Speed of water current required to transport sedimentary particles of varying grain sizes.

of these rocks are the limestones, consisting mainly of the mineral calcite (calcium carbonate: $CaCO_3$). This precipitation may take place directly from the water in warm shallow seas, such as the present-day situation in the Bahamas, or the calcite may be derived from the shells of marine animals who extract the calcite from the water, which is a form of biochemical precipitation. When the animal dies, the shell becomes available as a rock-forming component and such limestones may be referred to as bioclastic rocks, analogous to the clastic sediments already discussed.

Metamorphic rocks

The **metamorphic rocks** are simply any pre-existing rocks that have been changed (metamorphosed) by increasing pressure or temperature, or both. There are two broad categories: **regional** and **contact** (or thermal). Regional metamorphic rocks are produced by increased heat and pressure at great depths in the crust, and, as the name implies, this is a large-scale effect, often involving thousands of square kilometres. Contact metamorphism occurs when a rock, perhaps sediment, is in contact with a cooling igneous body. In this case the sediment is baked and hardened in much the same way as soft clay is transformed into hard brittle porcelain in the potter's kiln.

It is generally possible to separate regional and contact metamorphic rocks by significant textural differences. Most regional metamorphic rocks are foliated: they show a platy or wavy structure imparted to the rock by the parallel orientation of minerals such as the micas, which occur in the form of flat sheets like the pages of a book. This is epitomized by the slates, in which the foliation is so fine that thin sheets can be split from the rock and used as roof tiles. Contact metamorphic rocks tend to be granular in texture, for example **hornfels**, which is a fine-grain silicate rock of varied composition.

At subduction zones, such as under the Andes (Fig. 3.4), oceanic crust and sediments are pushed progressively deeper under the continental crust. This has the effect of increasing both pressure and temperature on the downward-moving plate. Any sedimentary rocks will be progressively changed by these increases in pressure and temperature, so that mudstone will be metamorphosed to **slate**, which also has a fine grain but possesses a very well developed foliation. Increasing metamorphism leads to a coarser-grain rock known as **schist**, containing flat-lying mica crystals in parallel orientation. At still higher levels of metamorphism, this orientation in schist becomes less pronounced and is replaced by medium to coarse bands of differing minerals and textures. This rock – near the limit of metamorphism and therefore just before the rock melts and becomes igneous – is referred to as **gneiss**. Further metamorphism could convert this gneiss to **migmatite**, which has both igneous and metamorphic characteristics and is considered transitional between metamorphic rocks and igneous rocks. Examples of the relationship between highly metamorphic rocks and igneous rocks derived from them will be examined in the north Donegal excursion (pp. 97–107).

Figure 3.8 shows the pressure/temperature conditions of regional metamorphism. The processes of weathering and the changes known as lithification – when, for example, sands change to sandstones – occur below about 300°C. The temperature required for the formation of gneiss is probably around 700° C and the pressure is the equivalent of burial at least 20 km deep in the crust. At temperatures and pressures above these values, the gneiss will change to migmatite, before melting to form a liquid that is granitic in composition. This granite magma will then rise towards the surface to form an igneous rock in the crust, which will then eventually be subjected to erosion and transportation, and the whole process of sedimentary rock to metamorphic rock will begin all over again (Fig. 3.4). This is referred to as the rock cycle.

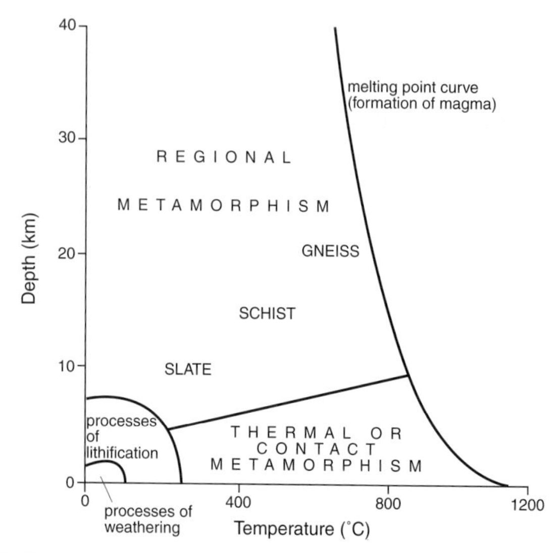

Figure 3.8 The pressure (as depth of burial) and temperature conditions of regional metamorphism.

Two non-foliated metamorphic rocks that are classified on the basis of their composition are **marble**, composed entirely of calcite and representing metamorphosed limestone, and **quartzite**, composed of quartz and formed by the recrystallization of sandstone. **Pelite** is the term often applied to fine-grain metamorphic rocks derived from sediments rich in clay minerals, such as shales and mudstones, and **psammite** refers to metamorphic rocks having the composition of sandstones.

Chapter 4

Geological time

No vestige of a beginning, no prospect of an end
James Hutton (1726–96)

The study of geology differs from that of most other sciences by including time in interpreting observations and drawing conclusions. The topics discussed in the Chapter 3 involve the concept of wholesale changes on the Earth's surface via the processes of plate tectonics and over very long periods of time. It is opportune now to consider the importance of understanding time in geology and the ways it can be measured. The objective of many geological investigations is to produce a geological history of an area: a sequence of events that can be built up from the evidence available in the rocks exposed. The quotation above is from the famous Scottish scientist who was among the first to realize the immensity of geological time. His further comment that "**the present is the key to the past**" recognized that the geological processes operating today are essentially the same processes, operating at the same rates, as those that have been active through the whole of geological time. This is the **principle of uniformitarianism**. As landscape modifications are so slow that during a typical human lifespan most changes will be imperceptible, Hutton realized that the time necessary to create the landmarks he was familiar with in Scotland could be measured only in millions of years. Precisely how many millions of years were involved was not calculated until well into the twentieth century, and the measurement of geological time is one of the great scientific achievements, involving contributions from not only geologists but also physicists and chemists. It is now thought that the age of the Earth, along with the Solar System, is about 4.5 billion years. The oldest rocks currently known on Earth are nearly 4 billion years old and the oldest rocks in Ireland, on Inishtrahull off the north coast of Donegal, are about 1.7 billion years old. Hominid evolution has occurred only in the past few millions of years, and the ice sheets of the most recent ice age, which were responsible for many familiar present-day landscape features, melted a mere 10 000 years ago. It is appropriate at this stage to examine the observations and techniques used by geologists to make sense out of the enormity of geological time and how ages such as those just quoted were arrived at.

The age of a rock can be measured in two ways. The **relative age** of a sedimentary rock is measured in comparison with other rocks in the same sequence: older than the rocks above it, but younger than the rocks below it. This is called the **law of superposition.**

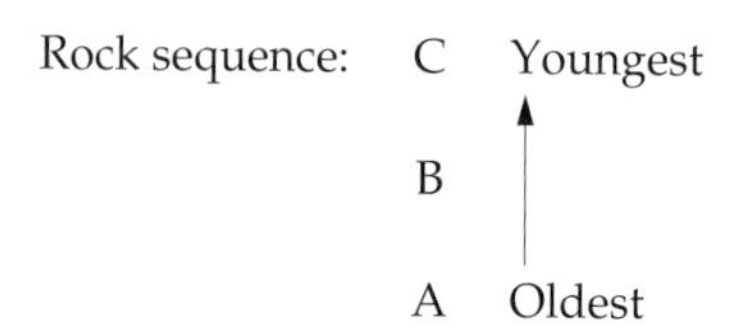

Rock B is older than C, but younger than A. The method can be refined if the rock sequences contain fossils. Similar fossil species, with some qualifications, can be used to imply similar ages of the rocks in which they are found, thus allowing correlation of rock sequences between different areas. This is the **principle of faunal succession**.

Two rock sequences, A,B,C and D,E,F, occur in two different areas. Within the sequences, rock B and rock F contain similar fossil ammonites. These can be used to say that rocks B and F are the same age, thus allowing the two sequences to be matched or correlated, even though they are geographically separated.

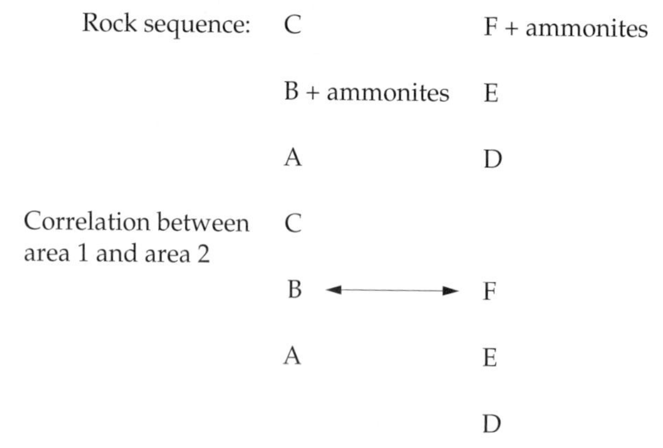

Rocks B and F are equivalent because they contain similar fossils, and rock D is the oldest rock in the two sequences.

Using this technique on a worldwide scale from the mid-nineteenth century onwards, it was possible to establish a geological or **stratigraphical column** based on the gradual evolution of life forms as shown by the fossil remains found in the rocks (Fig. 4.1). The subdivisions of geological time were defined by the general character of the life forms contained

Eon	Era	Period
Phanerozoic	Cainozoic	Quaternary
		Neogene
		Palaeogene
	Mesozoic	Cretaceous
		Jurassic
		Triassic
	Palaeozoic	Permian
		Carboniferous
		Devonian
		Silurian
		Ordovician
		Cambrian
Proterozoic		Precambrian

Figure 4.1 Stratigraphical column: main subdivisions of geological time.

within the rocks of that age. The largest subdivisions are the **eons**: **Proterozoic**, meaning "first life forms" and **Phanerozoic**, the eon of evident life. The Phanerozoic is divided into **eras**: the **Palaeozoic** ("ancient life"), the **Mesozoic** ("middle life") and the **Cainozoic** ("recent life"). These three eras are further divided into **periods**, based on the particular fossils present. The Cambrian is the oldest of the periods of the Palaeozoic, since it was mistakenly thought that life forms did not exist before this stage. Rocks older than Cambrian were designated as **Precambrian** and eventually it was realized that life extended far back into Precambrian times.

Stratigraphy is the description, correlation and classification of sedimentary rocks, including their environment of deposition. This method has worked particularly well for rocks younger than about 600 million years old, when abundant shelly fossils first appeared in sedimentary rocks. Geologists working with fossiliferous sediments have often been able to distinguish the relative ages of rock units only a few metres thick and representing time periods of only thousands of years' duration. In the mid-eighteenth century, the gradual acceptance of Charles Darwin's theory of evolution provided a framework for the use of fossils in a systematic way in the construction of a stratigraphical timescale. If it is assumed that life forms evolve from the simple to the more complex, then the evolution of the vertebrates followed the sequence:

fish—amphibians—reptiles—mammals.

If this sequence is placed in stratigraphical order, it can be represented as follows:

Youngest: Triassic fish, amphibia, reptiles, mammals (220 million years ago)
Permian reptiles (300 million years ago)
Carboniferous amphibia (370 million years ago)
Oldest: Devonian fish (400 million years ago).

This shows the evolutionary links between the main vertebrate groups between Devonian and Triassic times. The oldest rocks in this sequence will contain only fossil fish (Devonian), but the youngest (Triassic) will contain mammals and also fossils of the other three groups that have evolved from their common ancestors, the fish.

Thus, the geological column was established showing primitive life forms in the very oldest rocks, becoming progressively more complex through the Cambrian, Ordovician and Silurian, until, by the Devonian, fish were well developed and the first amphibians were beginning to colonize the land. The Carboniferous saw major advances in vegetation cover on land and the first reptiles appeared. By the Triassic and Jurassic, mammals had evolved, but life on land was dominated by dinosaurs, while ammonites were prolific in the oceans. The Cainozoic era contains many species that are similar to those currently in existence, including those that can be regarded as the precursors of hominids.

So, by the early twentieth century the relative age of most of the Earth's rocks had been established, but despite the great progress made in relative-age dating, no satisfactory method had been found at this stage to establish precisely how old rocks were. The second way in which the age of rocks can be measured is to calculate their **absolute age** (i.e. the age of the rock in millions of years). This followed the discovery of radioactivity at the end of the nineteenth century. Radioactivity is the natural production of radiation energy from radioactive elements such as uranium, and it occurs in certain minerals such as zircon, which is common in granites. Uranium is unstable and it breaks down or decays, giving out energy in the form of radiation and forming another element, lead. The changing proportions of the two elements can be seen in Figure 4.2, which shows the growth curve or increase in lead concentration with time, and the simultaneous decay curve or decrease of uranium concentration. In simple terms, by analyzing the amounts of uranium and its decay product (lead) in the rock and by calculating the rate at which this decay occurs, it is possible to calculate how long the process has been going on. In igneous rocks this is the time elapsed since the mineral crystallized, and in metamorphic rocks the time since the rock was changed and recrystallized. This is called **radiometric dating**, and by carrying out such calculations for very many samples all over the world, and combining them with the relative ages of the stratigraphical column, it has been possible to construct a more

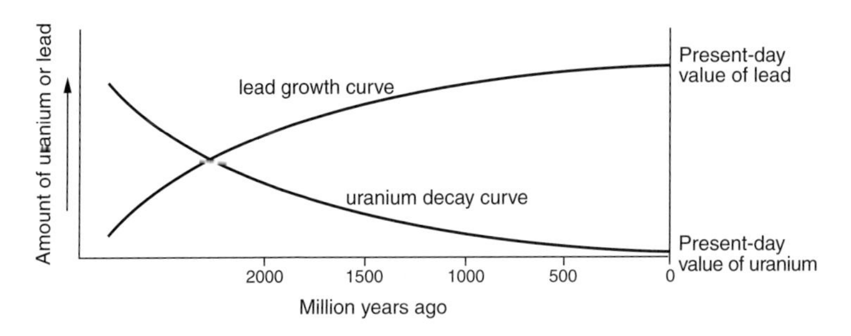

Figure 4.2 Growth/decay curve for lead/uranium, showing the changing proportions of the two elements with time.

comprehensive timescale for the rocks of the Earth. The stratigraphical column now can not only show rocks in sequence, but can do so within a precisely timed framework (Fig. 4.3). Thus, the age of the Earth was calculated at 4600 million years. The age of the Earth is calculated from measurements taken from meteorite specimens, which are considered to be the building blocks of the planets of the Solar System. One difficulty with the conventional display of the stratigraphical column is that it does not fully display the enormous length of Precambrian time (Fig. 4.4). The main part of a geological column is used to illustrate the subdivisions from the Cambrian to Recent, some 550 million years. Yet this ignores the previous 4000 million years of Precambrian time. Figure 4.5 attempts to redress the balance by showing geological time in the form of a spiral, with the origin of the Solar System and the oldest rocks on Earth at its centre. It is clear that evidence for geological processes becomes scarcer the further back in geological time one looks. The comparative wealth of information on such topics as evolution of plants and animals, plate-tectonic processes, and environments of deposition of sedimentary rocks that exists for the past 500 million years of geological time gives way to much less precise detail as the rocks get older, and more and more evidence is destroyed by the inexorable forces of erosion and crustal recycling.

Geological history: a case study

Figure 4.6 shows the sequence of rocks exposed on the east side of Murlough Bay near Ballycastle on the north Antrim coast. The exposed rocks are conspicuously different from each other, with the clifftop formed of the greyish-white limestone of the Cretaceous age Ulster White Limestone Formation, sometimes known as the Chalk. Underlying the limestone is a prominent red layer that consists of red sandstones and marls of Triassic

Age	Eon	Era	Period	Plate activity	Environment
1.6	PHANEROZOIC	Cainozoic	Quaternary	Himalayas formed; widening of N. Atlantic	Periodic advance and retreat of icecaps; glacially derived sands & boulder clay
23			Neogene	N. Atlantic forms by separation of Europe & N. America	Deposition of lignite and clays in lakes; lava plateau formed by fissure eruption
65			Palaeogene		
135		Mesozoic	Cretaceous	S. Atlantic widens; Gondwanaland fragments	Chalk sea covers much of Europe, including Ireland
205			Jurassic	Break-up of Gondwanaland continues	Marine conditions; mudstones with thin limestones and fossils
250			Triassic	Break-up of Pangaea; split of Gondwanaland and S. Atlantic opens	Arid continental conditions; sandstones and evaporites formed
290		Palaeozoic	Permian	Supercontinent Pangaea forms	Ireland in desert belt N. of Equator
350			Carboniferous	Ireland on Equator; Pangaea begins to coalesce	Marine or delta conditions; corals and coal deposits
410			Devonian	Caledonian mountain building; Ireland in desert belt S. of Equator	Deposition of red beds from rapid erosion of Caledonian Mountains; Donegal granites formed
440			Silurian	Continental collision, closure of Iapetus; Ireland at latitude of present W. Pacific	Greywackes formed in deep marine basin by slumps of material; island arcs formed; Dalradian sediments metamorphosed to schists and quartzites
510			Ordovician		
545			Cambrian	Opening of Iapetus	No examples in N. of Ireland; probably eroded
1000	PROTEROZOIC		Precambrian	Rifting of Rodinia about 750 million yr ago	Deposition of Dalradian sediments
2000				Formation of Rodinia about 1300–1000 million yr ago	Serpentinite of the Ox Mountains
3000				Ketilidian mountain-building episode 1780 million yr ago	Inishtrahull gneisses ~1780 million yr ago
4000		Archæan			Oldest known rocks on Earth ~4000 million yr old; formation of the Earth 4600 million yr ago
4600					

Figure 4.3 Stratigraphical column, showing absolute ages and the order of succession of the main rock types found in Ireland and their conditions of formation.

age. Below the steep cliff formed by these two rock units, the ground slopes much more gently and has an undulating hummocky appearance. This part of the slope is underlain by metamorphic rocks, which are mica-rich schists of Dalradian age.

Using a combination of absolute and relative ages for this sequence, it is possible to reconstruct the geological history for the area. When the approximate ages are added to the geological column, it becomes obvious that there are long gaps or breaks in the sequence (Fig. 4.7). The oldest

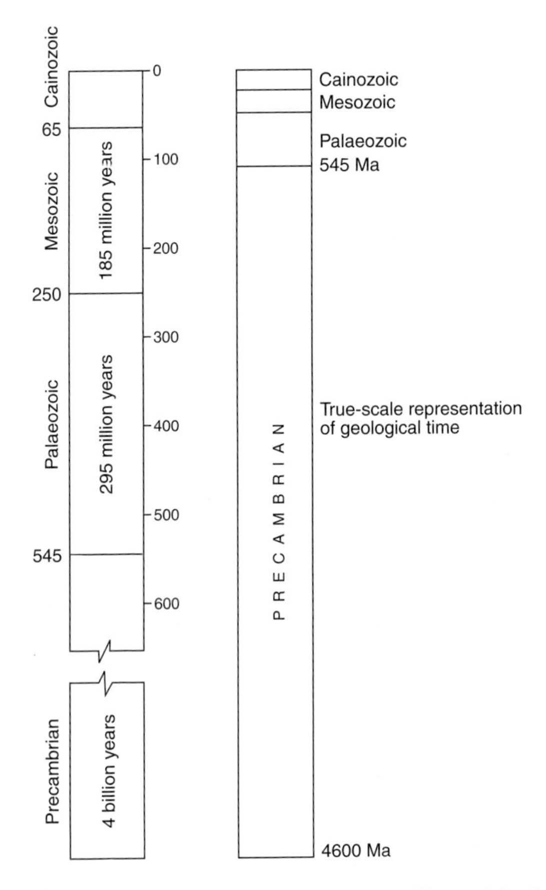

Figure 4.4 True-scale representation of the extent of Precambrian time.

rocks, the schists, are about 600 million years old, and lying directly on top of them are the Triassic rocks, approximately 250 million years old. This time gap, of about 350 million years, is called an **unconformity**. It is represented by a surface between two rock types that were not formed in a continuous or unbroken sequence. The layers missing between two such rock types were either never deposited or were removed by erosion before deposition of the younger rock. Similarly, between the red Triassic beds and the Cretaceous chalk there is a gap of about 75–100 million years, which is also an unconformity. Elsewhere in northeast Ireland there is a more complete sequence between the Triassic and the Cretaceous, including Jurassic rocks such as the Lias Clay. This suggests that the missing beds were simply eroded after deposition. Figure 4.7 also shows a sketch cross section of the Murlough Bay cliff with both unconformities marked.

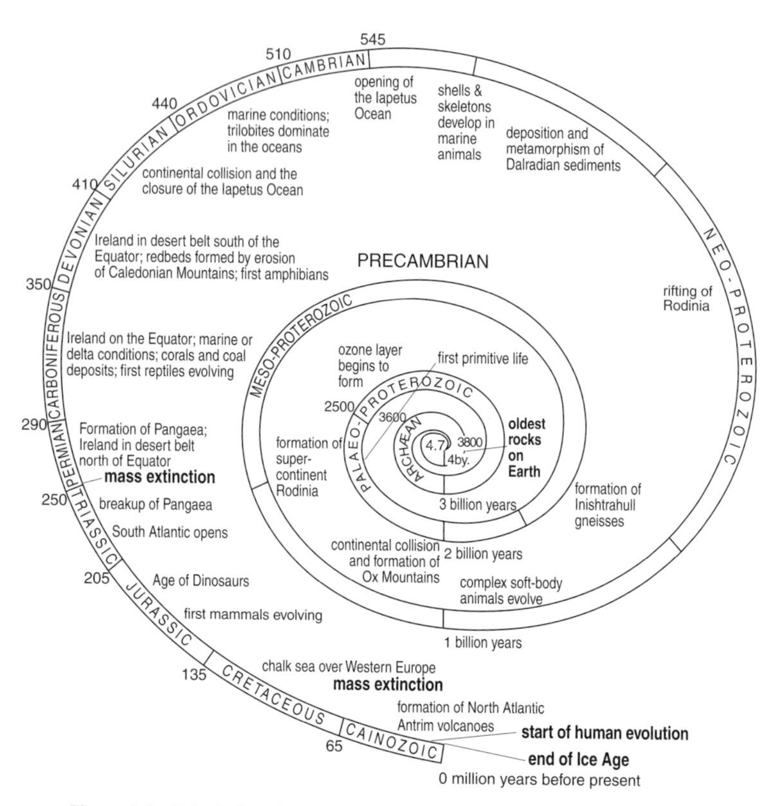

Figure 4.5 Extent of geological time from origin of the Solar System to present.

Figure 4.6 Sequence of rocks exposed in Murlough Bay, north Antrim.

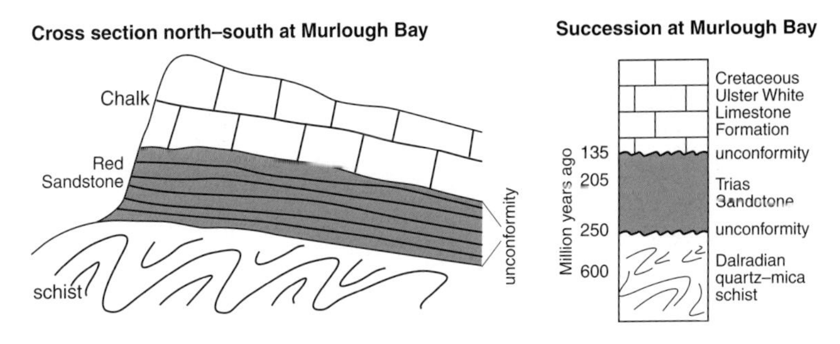

Figure 4.7 Cross section and succession, Murlough Bay, north Antrim, showing the major unconformities present.

The three rock types recognized here also provide evidence of markedly different environments of deposition, the details of which will be discussed in later chapters. The Dalradian sediments were laid down in intertidal or shallow marine conditions, before being folded and metamorphosed deep in the Earth's crust during a continental collision phase. The red Triassic rocks are the result of sedimentary processes operating in a hot continental desert environment, with the red staining caused by oxidation of the iron content. Ireland in Triassic times was north of the Equator, heading north and east towards its present position on the Earth's surface. Following deposition of the Triassic rocks there was another prolonged period of erosion that produced the second unconformity. The erosion surface was then drowned by the advance of a warm shallow sea in which countless billions of the microfossils known as **coccoliths** were deposited to form the White Ulster Limestone Formation. This sea covered virtually the whole of western Europe to the Ural Mountains in Russia, and in Britain its traces can be seen in the White Cliffs of Dover and as far north as the Isle of Skye.

The sequence of events shown by the succession can be interpreted as follows:

1. Deposition of the sandstones and mudstones of Dalradian age.
2. Folding and metamorphism of these beds in a mountain-building event.
3. Erosion of Dalradian rocks producing the unconformity surface.
4. Deposition of the Triassic sediments in desert conditions when Ireland was part of the supercontinent Pangaea.
5. Erosion of the Triassic sediments to produce the second unconformity.
6. Inundation of the land surface by a shallow sea to deposit the White Limestone.
7. Slight tilting and erosion to produce the present land surface.

Chapter 5

Geological history of Ireland

This chapter provides an outline geological history of Ireland in terms of plate tectonics, featuring the areas covered in the excursion guides.

The northern part of Ireland exhibits a great diversity of rock types, ranging in age from almost 2000 million years old to a few million years. These rock types were formed in a variety of geological environments, from hot deserts to icecaps, from active volcanoes to warm shallow seas containing corals, to deep marine basins in which vast thicknesses of sandstone and mudstone were deposited. In turn, these environments reflect the global position of Ireland throughout geological time as it responded to the forces of plate tectonics, which propel the crustal plates around the surface of the Earth. Reconstructing the former positions of crustal segments through time is known as palaeogeography. In general terms, an indication of the former position of a crustal segment can be established from the rock types included in that segment. For example, desert sandstones are generally limited to the arid regions immediately north and south of the Equator, whereas glacial sediments are restricted in the main to the regions around the North and South Poles. Limestones occur in the equatorial and subtropical belts, and coal deposits require the high growth rates of the equatorial or the warm temperate belts.

Using features of the Earth's magnetic field that are often preserved within certain rocks at the time of their formation, it is often possible to calculate the former positions of crustal segments on the surface of the Earth. The Earth's magnetic field, or geomagnetic field, can be measured at any point on the Earth's surface. A compass needle swings and points to magnetic north, and the angle between geographical north and magnetic north is the angle of **declination** (Fig. 5.1a). The needle also points down into the Earth and, the nearer to the magnetic pole, the steeper the angle at which the needle dips. The angle the needle makes with the horizontal is the angle of **inclination** (Fig. 5.1b). This feature of the geomagnetic field can therefore be used to calculate how far away the geomagnetic pole is, which shows how the angle of dip relates to the distance from the magnetic pole (Fig. 5.1c). The declination gives the direction to the magnetic pole. A

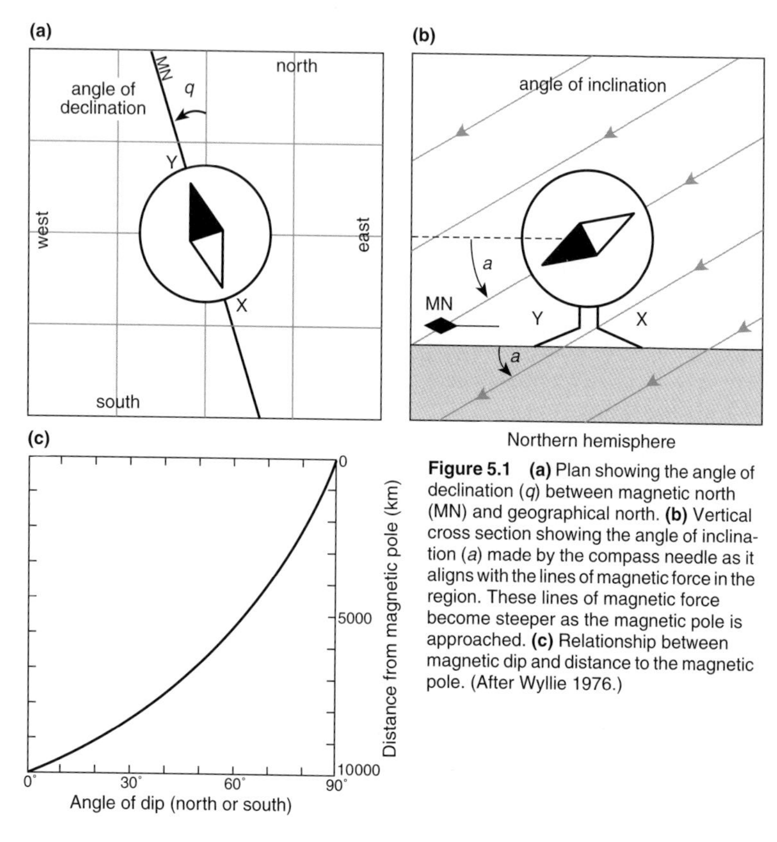

Figure 5.1 **(a)** Plan showing the angle of declination (*q*) between magnetic north (MN) and geographical north. **(b)** Vertical cross section showing the angle of inclination (*a*) made by the compass needle as it aligns with the lines of magnetic force in the region. These lines of magnetic force become steeper as the magnetic pole is approached. **(c)** Relationship between magnetic dip and distance to the magnetic pole. (After Wyllie 1976.)

naturally occurring magnetic mineral known as magnetite occurs in many rocks, especially basalts. As the basalt cools after eruption, the crystals of magnetite align themselves with the magnetic field acting on them at that time, just as if they were miniature compass needles. This provides an imprint of the magnetic field at the time and place of origin of the lava, and by using this "fossil" or remnant magnetism it is possible to work out the distance to the geomagnetic pole and its direction at that time. From that information the approximate position on the Earth's surface where the lava originated can be estimated. The problem here is that, although values for latitude may be calculated, there is no constraint on the longitude, and in theory each crustal fragment could lie anywhere on the line of latitude calculated. However, despite these limitations, a general consensus exists that allows broad agreement on the path taken by Britain and Ireland over the past 1000 million years or so. Figure 5.2 shows the estimated change in the position of Ireland from the late **Ordovician period** (about 450 million years ago) to the present day. Ireland at that time was

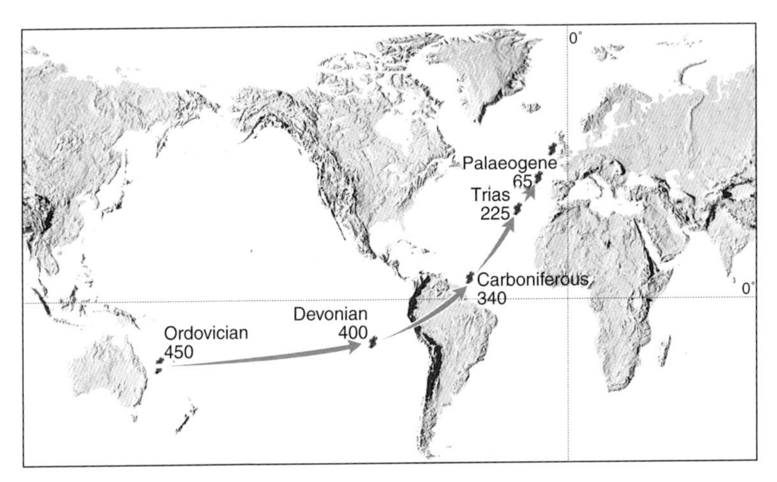

Figure 5.2 Estimated global position of Ireland from the late Ordovician to present.

south of the Equator, somewhere in the present-day western Pacific, and beginning a long journey, generally eastwards and northwards to its present position. During this odyssey, Ireland experienced climatic changes as latitude lines were crossed and it passed through hot desert zones, on either side of the humid equatorial zone, to its present position in the cool temperate zone. The red desert sandstones of the **Devonian** period (about 400 million years ago) and the **Triassic** period (about 250 million years ago), which bracket the coral-rich limestones of the **Carboniferous** period (about 350 million years ago) reflect these changes in latitude with time (see Fig. 4.2).

Throughout geological history, continents have existed in two patterns: either dispersed over the surface of the Earth or nucleated into one or two supercontinents. Periods of supercontinent formation were characterized by low rates of **seafloor spreading** and therefore low volumes of mid-ocean ridges. Conversely, continental break-up meant high rates of seafloor spreading and therefore high volumes of mid-ocean ridges. These increased volumes of submarine ridges meant that sea levels would be pushed up and, when ridge volume was low, there would be a corresponding lowering of sea levels globally (Fig. 5.3). These periodic changes in sea levels were important influences on the geological environment, particularly as changes in the rates of seafloor spreading meant there were changes in carbon dioxide emissions from volcanoes into the atmosphere. Increased volcanic activity meant higher carbon-dioxide levels in the atmosphere and therefore a warmer world, since increased CO_2 levels cause greater heat retention in the atmosphere. This is now known as global warming, which is currently increasing because of the burning of fossil

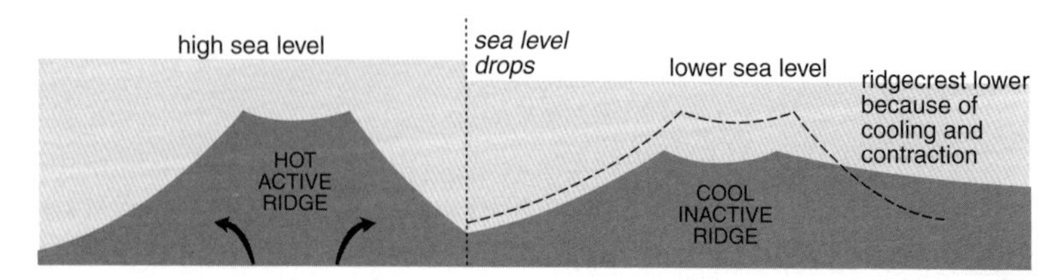

Figure 5.3 The relationship between sea levels and submarine-ridge volumes. During active spreading periods, ridge volumes are high, which would push up sea levels, whereas low rates of seafloor spreading would mean lower ridge crests because of cooling and contraction, and therefore correspondingly low sea levels.

fuels, but which has been varying naturally throughout geological time. Conversely, when CO_2 levels were lower, the world was correspondingly cooler. At the same time as all these changes were occurring, biological evolution was taking place, as increasingly complex life forms developed and moved from the oceans to colonize the land areas. This overall tendency for increased biological diversity has been punctuated by the periodic extinction of many species, of which the events at the end of the **Permian** and **Cretaceous** periods were the most extensive. The geological history of Ireland can therefore be written in terms of changing environments through time while Ireland drifted over the surface of the Earth as a component part of various continental plates. This was a complex pattern of movement that saw the building of supercontinents by collision or accretion of continents, followed by the eventual fragmentation or break-up of these supercontinents, before the cycle was repeated and the continents re-assembled. Supercontinent formation took place via a series of continental collisions that resulted in orogenies or mountain-building phases. In the same way that the head-on collision of two cars tends to crumple and lift the bonnets of the vehicles involved, so crustal collision is characterized by compression or shortening of the crust and simultaneous vertical uplift. The boundaries of these crustal collisions often mark zones of major geological change, and in Ireland one of the most conspicuous is the zone that marks the **Caledonian orogeny**. From late **Silurian** to early Devonian times, this welded together the continental masses of Laurentia and Avalonia along a line known as the **Iapetus suture**. This continental collision laid the foundations of the island of Ireland as it now exists, and it produced the Caledonian Mountains, which were as high as the present-day Himalayas and whose remnants are seen now in the highlands of Norway, Scotland, Donegal and northeast Canada. By convention, geologists work from the oldest event to the youngest, so to interpret the geological history of the north of Ireland it is necessary to go back to the beginning, or as near the beginning as possible.

The Precambrian

There is direct evidence in the rocks of the north of Ireland for plate-tectonic processes happening perhaps as far back as 1700 million years ago. However, since the age of the Earth is estimated at 4500 million years, and the oldest rocks known on Earth are almost 4000 million years old, then it is obvious that there are still huge gaps in our complete understanding of the entire geological history of Ireland. These gaps may never be satisfactorily filled, as all of the rocks from these early stages of Ireland's development, and therefore all of the evidence, may well have been destroyed by erosion.

The oldest rocks that occur in the part of Ireland covered by the 1:250 000 map are the coarse metamorphic rocks known as gneisses, which occur on the island of Inishtrahull, about 10 km off the north coast of Donegal. It is thought that rocks similar to these underlie the younger rocks exposed on the mainland. They represent an episode of mountain building that took place about 1700 million years ago and for which very little evidence remains. The Ox Mountains to the southwest and northeast of Sligo are formed of some of the oldest rocks on the Irish mainland. They are metamorphic rocks, including schists and gneisses, and were deposited as shallow marine sediments more than 1000 million years ago. Ireland was then part of a continent comprising North America and Europe, and was situated somewhere near the South Pole. By about 1000 million years ago, plate movement had produced a single supercontinent on the Earth's surface, called Rodinia. This consisted of the continental fragments Laurentia (North America, Greenland), Siberia and Baltica (Scandinavia, the Baltic countries) and the various continental masses known as Gondwanaland (India, Antarctica, Australia, South America, Africa). The continental collisions involved in this accretion pushed the marine sandstones deep into the Earth's crust, where they were metamorphosed into the schists and gneisses of the Ox Mountains. By 600 million years ago this supercontinent appears to have fragmented and re-formed as a second supercontinent, the Vendian, comprising Gondwana, Laurentia and Baltica. Figure 5.4 shows the Vendian supercontinent at approximately 600 million years ago, just before the beginning of fragmentation as it split again to form the distinct continents, Laurentia and Gondwana, separated by a new and expanding ocean. Quartz sands, muds and lime muds were deposited in this ocean in a sequence that indicated deepening water. These sediments provided the basis for the most widespread bedrock in Donegal, the **Dalradian metasediments.**

The **Dalradian Supergroup** comprises a belt of metamorphic rocks that stretches from Connemara north towards Donegal, Tyrone and northeast

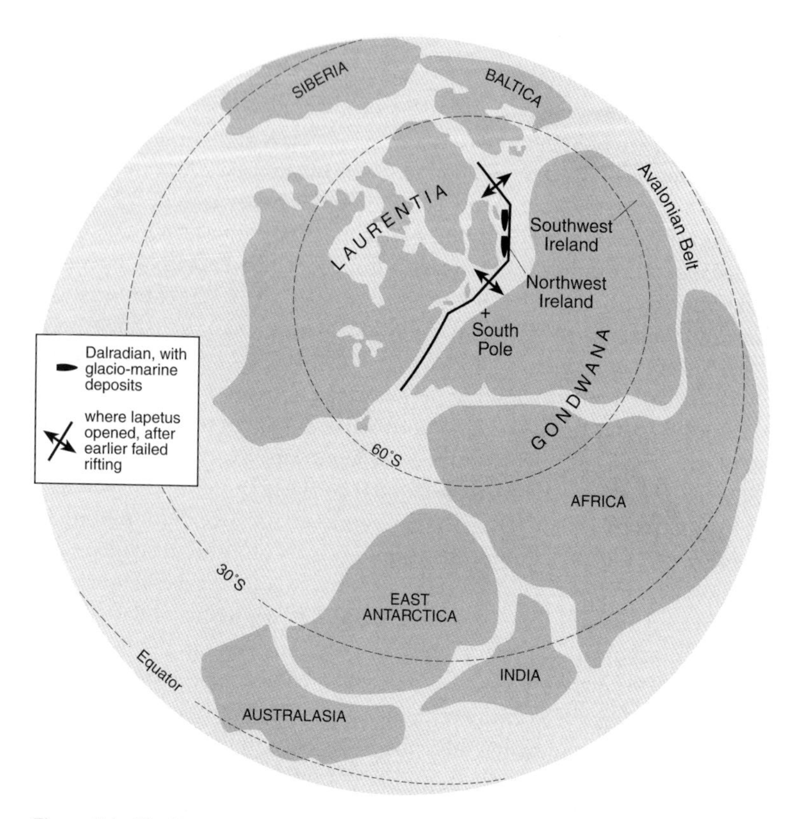

Figure 5.4 The Vendian supercontinent at approximately 600 million years ago (after Long & McConnell 1999).

Antrim and into the Highlands of Scotland. They are mainly intertidal and marine sediments, but they include various intrusive and extrusive igneous rocks and some glacially derived sediments. The sediments were recrystallized and converted into metamorphic rocks during a mountain-building phase during the later Ordovician period.

The Cambrian, Ordovician and Silurian

The fragmentation of Rodinia and the Vendian supercontinent produced two smaller continental blocks: Laurentia (North America, northern Europe, Siberia) and Gondwana (Africa, southern Europe, South America, India, Australasia). The further rifting of Gondwana produced the break-away small continental mass of Avalonia, which contained the southern portion of Britain and Ireland (Fig. 5.5). The northern parts of Britain and

Ireland were included in Laurentia. This meant that, in Cambrian times (about 500 million years ago), what is now recognized as Ireland was split in two parts on either side of a widening ocean known as the Iapetus (Greek mythology: Iapetus was the father of Atlas, after whom the Atlantic Ocean is named), situated south of the Equator (Fig. 5.6a). This ocean was a sort of pre-Atlantic, as it separated a continental plate consisting mostly of North America and another recognizable as predominantly Europe. Deposition of the Dalradian rocks came to an end with the spreading open of the Iapetus Ocean during the Cambrian period. Around the end of Cambrian times, plate movements changed again and the ocean began to contract during the Ordovician period. Remnants of the deep marine sediments deposited in the Iapetus Ocean during Ordovician and Silurian times form the extensive belt of sandstones, shales and mudstones that are found in a southeast-trending zone from County Down on the east coast through Armagh and into Monaghan and Cavan. They are primarily the

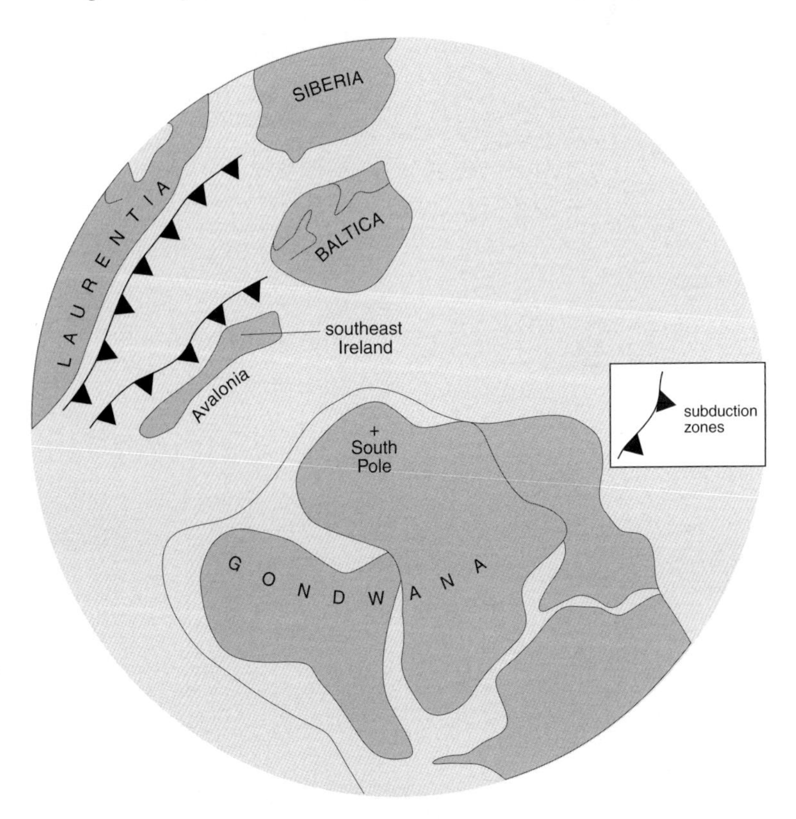

Figure 5.5 The break-up of Rodinia and the Vendian supercontinent approximately 470 million years ago (after Woodcock & Strachan 2000).

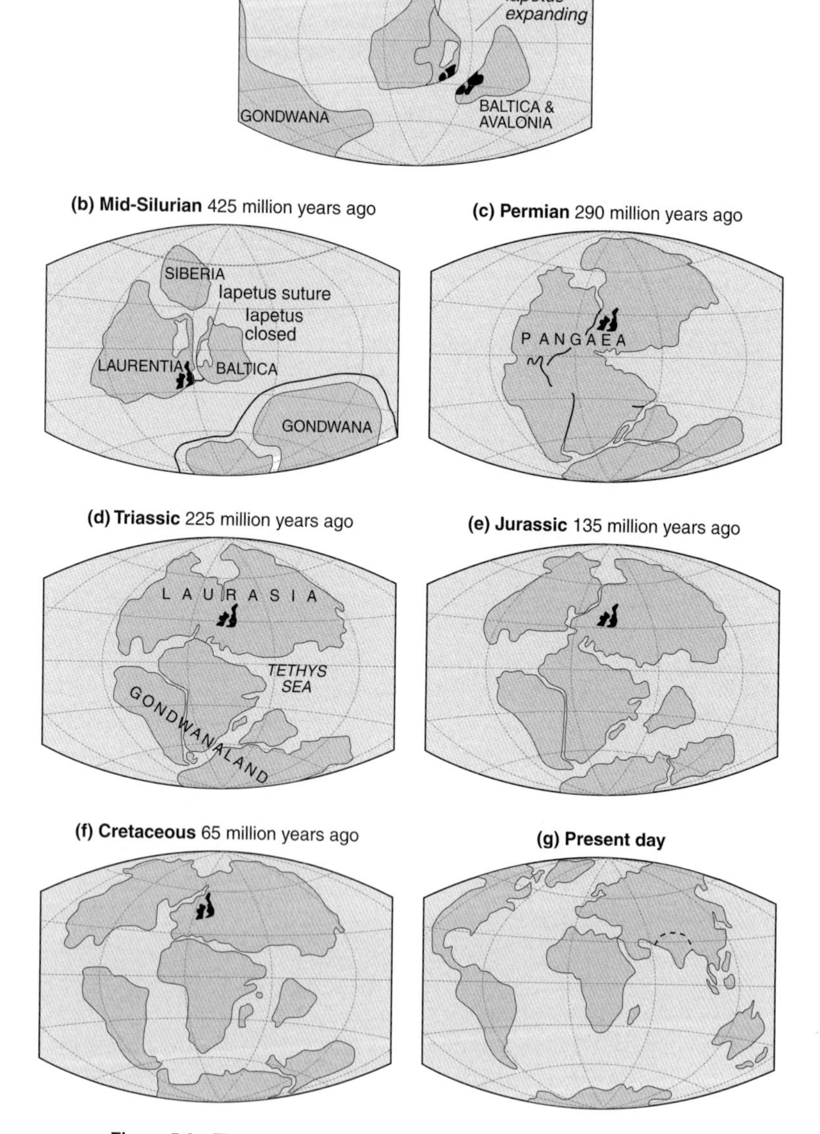

Figure 5.6 The global distribution of continents from Cambrian to Recent.

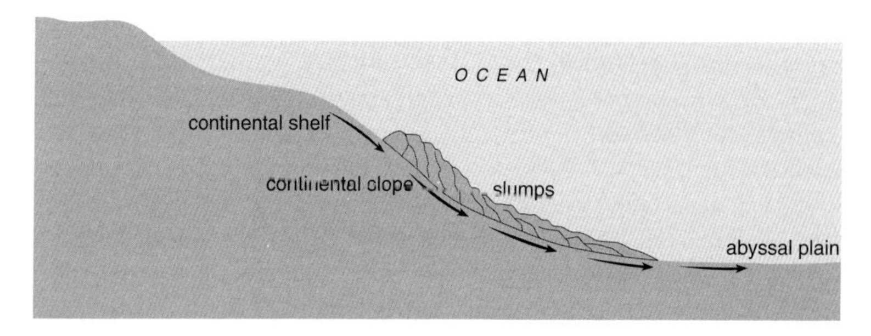

Figure 5.7 Formation of turbidites: slumps on slope, triggered by earthquake, generate turbidity currents that flow down continental slope to abyssal plains, where they are deposited as turbidites

result of a phenomenon known as **turbidity currents**, which occur when accumulations of sediments on the continental shelves become unstable and slip down the continental slope as a dense current of mixed sediment. When the flatter ocean floor (the abyssal plain) is reached, the current loses momentum and deposits the sediment load as beds of sand, silt and clay, often graded from coarse at the bottom to fine upwards, which are referred to as **turbidites** (Fig. 5.7). Their marine origin is indicated by their fossil content and there is evidence of contemporaneous igneous activity in the form of **pillow lavas** (which occur when eruptions take place under water) and thin volcanic ashbeds. Localities on and around the Ards Peninsula in County Down will be used to illustrate these features.

As the two continental masses converged because of the closure of the ocean, the resulting continental collisions altered the older shallow-water marine mudstones, sandstones and limestones into the schists, quartzites and marbles of the Dalradian Supergroup. The horizontal compressive forces of the mountain-building process folded, fractured and deformed these originally near-horizontal sediments, and in many areas of Donegal complex folds are visible. Exposures at Horn Head in north Donegal show evidence of the large-scale movement of blocks of rock along **thrust planes**, leaving older schists lying directly above younger paler quartzite.

The deep-water marine sediments that formed in the Iapetus Ocean and are exposed so well on the coast of County Down were also folded and fractured during this phase of continental collision. The pillow lavas and ashes included in this sequence were probably associated with a series of **island-arc** volcanoes that formed as part of the subduction processes at the northern margin of the Iapetus Ocean. These volcanoes were similar to those currently active in the Philippines.

Continuation of the compressive forces involved in continental collision and the closure of the Iapetus Ocean, initiated a prolonged period of mountain building involving a complex series of continental and island-arc collisions that formed the Caledonian Mountains. This mountain range not only included areas of Ireland and Scotland but also extended northwards into Norway and southwards into Newfoundland and Nova Scotia in Canada. Figure 5.8 shows the details of the collision of Laurentia with Avalonia and Baltica as the Iapetus Ocean finally closed, producing the Caledonian Mountains or Caledonides. These mountains were at times on a scale comparable to the present-day Himalayas and they dominated the geology of the north of Ireland until the Carboniferous about 75 million years later. The Iapetus Ocean opened in the Cambrian, reached a maximum width in the Ordovician and ended finally with closure in the early Silurian, a total period of about 100 million years. An important result of the closure of the Iapetus Ocean and the prolonged mountain-building phase was to bring the two parts of Ireland together and put in place the framework on which the succeeding geological structure of the island was based (Fig. 5.6b).

The closure of the Iapetus Ocean was a tectonically complex series of collisions involving not just the main continental masses of Laurentia, Gondwanaland and Baltica, but also fragments of Gondwanaland, such as Avalonia, which contained southern Britain and the south of Ireland. These collisions are imprinted on the crustal fabric of Britain and Ireland,

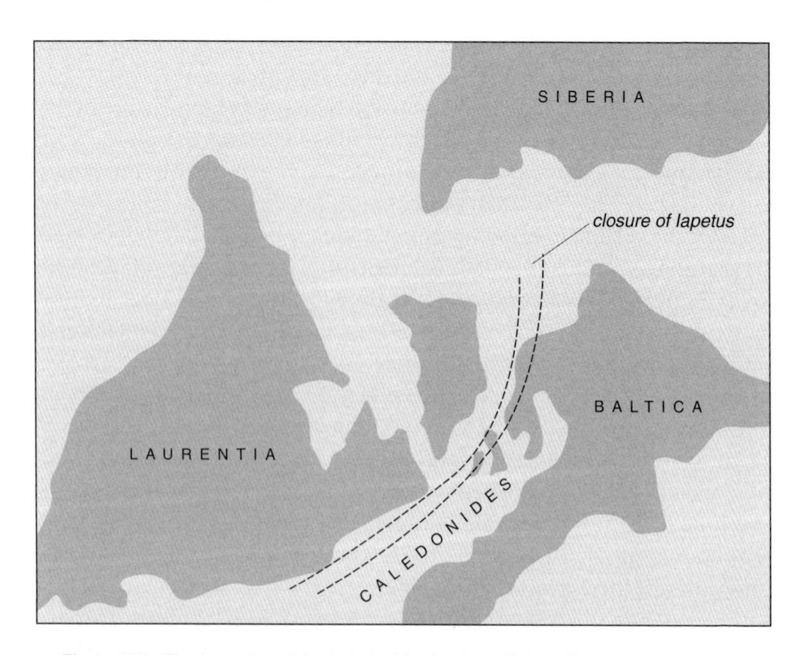

Figure 5.8 The formation of the Caledonides by the collision of Laurentia and Baltica.

often as major fault lines or fold belts. In the case of the Caledonian orogeny, the dominant trend or grain of this imprint is northeast–southwest. In places across Ireland and Britain, the join or suture between the two previously separate continental fragments can be seen. These newly joined continents have distinct geological histories, are fault-bounded and are referred to as **terranes**. They are displaced or exotic fragments of continent or island arc or ocean basin, which have been plastered or accreted onto continental margins by collision and subduction during the process of continental break-up and re-assembly. In Ireland the Iapetus suture is thought to run from the Shannon Estuary in the southwest, through to Clogher Head, north of Drogheda, on the east coast. It then crosses the Irish Sea, beyond the most northerly point of the Isle of Man, to the Solway Firth, and crosses England north of the Lake District and south of the Scottish border. The terranes in northwest Ireland, north of the Fair Head–Clew Bay line, which is considered to be a continuation of the Highland Boundary Fault, are thought to show Laurentian affinities. Those south of the Iapetus suture have affinities with Gondwana. Between the two is an intermediate zone consisting of island-arc and oceanic fragments, such as will be examined in the Tyrone excursion (pp. 117–126). The Orlock Bridge Fault in north County Down is a likely terrane boundary within this intermediate zone (Fig. 5.9).

The result of the continental collision that closed the Iapetus Ocean can still be clearly seen in the often prominent southwest–northeast structure or grain of the north of Ireland. A glance at the geological map will show this most clearly in Donegal. Here the Main Donegal Granite, which forms the Derryveagh Mountains, and many of the principal fault lines such as the Gweebarra Fault or the Leannan Fault, are aligned in this direction, often referred to as the Caledonian trend.

The metamorphic rocks of the Dalradian Supergroup make up much of the bedrock of central and north Donegal, but the other dominant rock type is granite. The tectonic forces involved in subduction and continental collision, which were responsible for metamorphosing the Dalradian rocks and throwing up the Caledonian Mountains, were also generating sufficient heat at great depths in the crust to cause crustal melting (see Fig. 4.3).

As is the case in modern-day mountain chains such as the Himalayas, the pressures and temperatures involved are sufficient to take metamorphic processes beyond the limits of solid-state change to where the rocks melt. This produces magma of broadly granitic composition at depths of about 30–40 km in the crust. This melted liquid is less dense than the surrounding unmelted country rock and it will tend to rise in the way that oil globules will rise through relatively dense water. In subduction or collision zones, the granite magma body will seek out lines of weakness

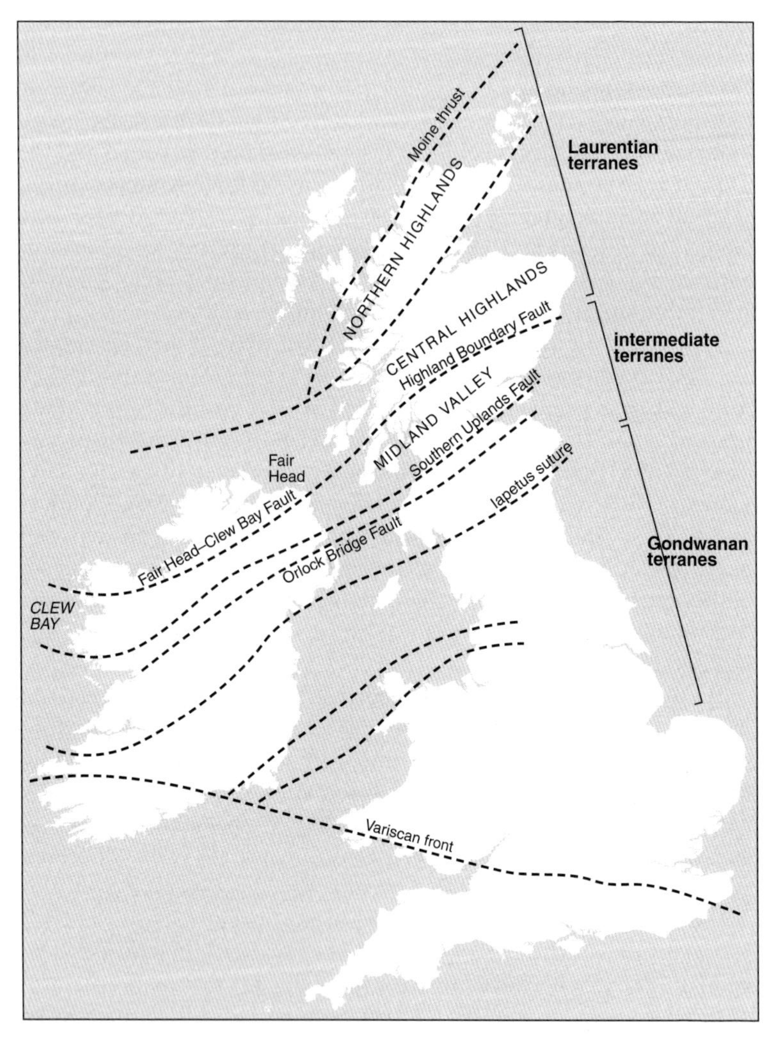

Figure 5.9 Simplified fault lines and terrane boundaries in Britain and Ireland (after Woodcock & Strachan 2000).

and rise to a region in the crust where the density of the solid crust matches the magma, so that the magma body will cool and solidify. Much of the Donegal Highlands are formed of granite masses (e.g. the Derryveagh Mountains), as typified by the spectacular glaciated scenery around the Poisoned Glen at Dunlewy. Granite intrusion continued until about 400 million years ago, well into the succeeding Devonian period.

The Devonian

With the disappearance of the Iapetus Ocean and the formation of the vast Caledonian mountain chain, the scene was now set for a complete change of geological environment for Ireland. At this time (about 400 million years ago), Ireland was on the southern margins of the newly formed Laurentian continent, the ocean being at some distance to the south. Ireland was still located in tropical latitudes south of the Equator and was continuing on the generally eastward and northward drift that had proceeded since the closure of the Iapetus Ocean (Fig. 5.2).

Conditions changed from predominantly marine to a continental environment, because of the rainshadow effect of the mountains to the north, specifically to continental desert conditions. Rainshadow occurs when the prevailing winds blow across the mountains and drop most of their moisture as rain or snow on the high ground, leaving little rainfall for the adjacent plains. The high Caledonian Mountains to the north would have attracted much of the rain from the prevailing winds, leaving little to fall on the lower ground to the south. This rainshadow effect is clearly demonstrated today in areas such as Death Valley in California, which is in the rainshadow of the Sierra Nevada mountain range to the west. In the absence of any protective vegetation cover (land plants were not yet well developed), these mountains would have been very vulnerable to erosion by wind and water, and huge thicknesses of sediments were deposited at the foot of the mountains. In Death Valley, torrential streams (flashfloods) carry large quantities of unsorted sedimentary debris and dump them in fan-shape deposits at the base of the mountains. The fragments in these deposits have often been rounded by collision with each other in the streams and are deposited as conglomerates. These conglomerates are predominantly formed of the harder more resistant rock types such as quartzites or granites, whereas softer rocks such as shales or limestones have broken down into finer particles found as sand or clay particles in the matrix of the conglomerate.

Sediments of this age, often referred to as Old Red Sandstone, are Devonian in age and are found in the Clogher Valley in County Tyrone, on the Fanad Peninsula in Donegal and, most notably, around Cushendall and Cushendun on the east coast of Antrim. The Devonian period, as well as being characterized by hot continental desert conditions, is also noted for the continuation of plant colonization on land and the evolution of many forms of fish in the oceans, including sharks and the earliest bony fish. A further significant development in Devonian times was the evolution of amphibians and therefore the beginning of colonization of the land by animal life.

The Carboniferous

The end of the Devonian, at about 350 million years ago, was marked by a gradual flooding of the desert landscape by a sea that advanced from the southeast as the Earth passed into one of its higher sea-level phases. The succeeding Carboniferous period is characterized by a series of advances and retreats of this sea. The changes from terrestrial to marine conditions and back again is a theme repeated constantly throughout geological time and they are part of the global pattern of marine fluctuations related to plate-tectonic processes described earlier. The continental sedimentation that dominated the Devonian was the result of the Caledonian collisions that formed the enlarged continent Laurentia and the Caledonian Mountains, leading to lower sea levels. The early Carboniferous transgression resulted from increased plate-spreading activity, and therefore higher sea levels, that occurred between the formation of the supercontinent Laurentia and the later supercontinent of **Pangaea**, which began to coalesce in the middle to late Carboniferous.

The erosion of the Caledonian mountain ranges was by now well advanced and the low-lying desert plains that constituted Ireland at that time were drowned by the encroachment northwards of a warm shallow sea. In this sea lived a wide range of animal life, including many species of fish, shellfish and corals. Sediments of Carboniferous age, particularly limestones, are the most common bedrock in Ireland.

The earliest deposits from this advancing sea were conglomerates and sandstones, but, as the water became deeper, these soon gave way to finer sandstones, siltstones and eventually limestones. Limestone is predominantly the mineral calcite, which forms either by chemical precipitation from sea water in warm shallow seas, as is currently taking place in the Bahamas, or is produced biologically by shellfish or corals. When the animal eventually dies, the calcite remains as a potential component of limestone. A feature of the Carboniferous in Ireland was that sea level appeared to fluctuate, so that deepwater limestone formation changed to shallower-water sedimentation in deltas, before reverting back to limestones as sea level rose again. When the area was close to the ancient shoreline as indicated by delta formation, then sandstones were typically produced. Ireland was now very close to the Equator (see Fig. 5.2). Luxuriant plant life had developed on land, and the Carboniferous period was marked by the development of huge damp forests with thick mangrove-type swamps around the coast. High growth rates in the hot moist conditions meant that much dead vegetation accumulated very quickly, often in stagnant water with little oxygen, which inhibited its decay and disintegration. As more and more vegetation was deposited, the pressure increased and the plant remains were compressed, gas and water were driven out, and coal of various types formed. Although Ireland lacks the extensive coalfields found in Britain in rocks of similar age, areas such as Coalisland in County Tyrone and Ballycastle in County Antrim had enough thin coal seams to have supported a small-scale coal industry until the 1950s. Surviving traces of the exploitation of the north Antrim coalfield will be examined in the Ballycastle excursion (pp. 169–177). Offshore from the deltas the deeper water allowed the formation of limestones from lime mud and shell fragments, such deposits often being topped by coral reefs. Abundant fossil corals are found in Fermanagh and Sligo in localities such as Streedagh Point, County Sligo.

Karst scenery

Calcite is a relatively soft mineral (softer than steel), unlike quartz, which can scratch steel. More importantly, it dissolves in rain water or acidic groundwater. This leads to a characteristic weathering process in limestone areas – referred to as karst scenery after an area in the former Yugoslavia where it is particularly well developed – where erosion is dominated by solution of the country rock (Fig. 5.10). Rain water and groundwater percolate down through cracks and joints in the limestone, widening them by dissolving the surrounding calcite grains. When the percolating water is prevented from moving further downwards by an impermeable layer of rock such as shale, it begins to move laterally, dissolving limestone all the while. The enlarged vertical cracks or joints become **sinkholes**, the laterally moving streams of water dissolve out channels that eventually enlarge into galleries and caves, and a complex underground drainage system is developed as the surface streams disappear down the sinkholes. These features are spectacularly illustrated in the show caves at Marble Arch in the Carboniferous limestone of County Fermanagh (pp. 109–116). The characteristic hill shapes of west Fermanagh and Sligo, as typified by the flat-top hill of Benbulben, are explained

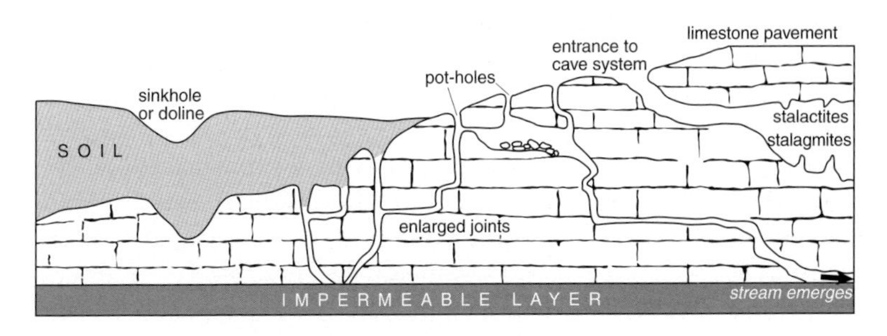

Figure 5.10 The characteristic landforms of a karst area.

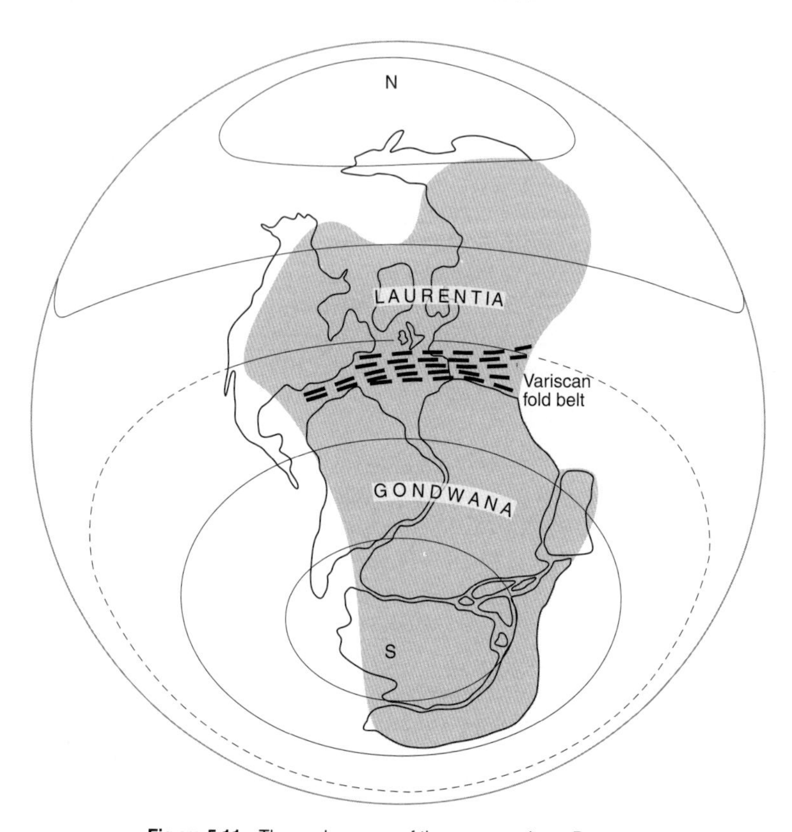

Figure 5.11 The coalescence of the supercontinent Pangaea.

in terms of the mainly horizontal beds of limestone being protected from erosion by an overlying harder and more resistant sandstone bed. These nearly horizontal limestone and sandstone beds reflect the fact that no mountain-building episodes since the Caledonian orogeny have impacted on the rocks of the north of Ireland. By the middle of the Carboniferous the continental masses on the Earth's surface were again moving together to form the next supercontinent. Called Pangaea (Greek: "all land") and surrounded by a single ocean, it was to form a single continental block running north–south (Fig. 5.11). The collision of Laurentia and Gondwana towards the end of the Carboniferous produced a mountain-building episode, the Variscan orogeny, which produced downfolded and downfaulted blocks of sediments that preserved the extensive Carboniferous coal deposits of England and Wales. As Ireland was mostly unaffected by the Variscan orogeny and lacks these preserved crustal blocks, any extensive coal beds formed in Ireland in the Carboniferous were subsequently eroded away.

As happened also in Devonian times, this phase of continental collisions and accretions increased the overall volume of the ocean basins because of a reduction in seafloor spreading and therefore a reduction in the size of the mid-ocean ridges. This lowered sea levels and raised erosion rates, so that once again large thicknesses of eroded material were deposited and hot desert conditions returned, marking the beginning of the Permian period in Ireland.

The Permian

The formation of the great continent Pangaea in Permian times brought about a radical change in environment for Ireland (see Fig. 5.6c). The marine or deltaic conditions that had prevailed during the Carboniferous were replaced by hot desert conditions, as Ireland, now firmly held in a vast continental interior, drifted north of the Equator. Conditions were not unlike those of the earlier Devonian period, with the formation of reddish sedimentary rocks. These included rocks composed of angular fragments (known as **breccias**), and also windblown dune sands. In these, the sand grains are characteristically very well rounded by the constant collisions with other grains as they were transported by strong winds. The breccias can be seen at Cultra on the coast of north Down. A marine advance (the Bakevellia Sea) from the north covered parts of northeast Ireland in later Permian times, accounting for the fossiliferous marine limestones of Permian age found in north Down near Cultra.

At about 250 million years ago, the Bakevellia Sea retreated and a major catastrophic event (or events) occurred, which produced a mass extinction of species greater than anything previously experienced on Earth. This marked the end of the Permian and also the Palaeozoic era. The cause or causes of this extinction remain unclear. Unlike the later mass extinction that accounted for the dinosaurs and other species at the end of the Cretaceous period nearly 200 million years later, there is no compelling evidence for an asteroid or comet impact. The unique distribution of the continents in a single supercontinent may well be an important factor. Many of the species that died out were marine, so an explanation must be sought that involves changes in the marine environment. Evidence exists for a phase of global cooling and a contemporaneous drastic reduction in marine oxygen levels, and these could account for at least the marine part of the extinction.

Whatever the causes, the extinction event at the end of the Permian marked a turning point for life on Earth. Important groups such as the trilobites, which had existed since early Cambrian times, were never seen again, even though the last Permian examples appeared to be thriving.

None of the previously existing coral species survived and the coral species that replaced them were different in symmetry and in the mineral composition of their skeletons.

The Triassic

The period following the Permian is the Triassic, the first period of the Mesozoic, and is known as the Age of Reptiles because of the eventual domination of life forms by the dinosaurs.

The beginning of the Triassic in Ireland was marked by a return to desert conditions. Ireland was still part of Pangaea, lying at about the latitude of the present-day Sahara and separated from the ocean by the Variscan mountain chain, which continued to act as a sediment source (see Figs 5.2, 5.6d). The sediments deposited in the earlier part of the Triassic were mainly sandstones from rivers that crossed the generally low-lying plains. Mud layers showing ripples and cracks preserved in these sandstones indicate shallow-water conditions with occasional periods of drying out. Some of these sandstone beds were re-worked by the wind to form dunes. The exposures in the former quarries around Scrabo Hill near Newtownards provide the best examples of these rocks in Ireland, which were used as a building stone for many years in the nearby town and also farther afield in Belfast and Dublin. The Scrabo sandstones, known formally as the Sherwood Sandstone Group, have a high porosity and are important water-bearing rocks, or **aquifers**, for Newtownards and the Lagan Valley area around Belfast. The long-established mineral-water industry in Belfast is based on the ability of these sandstones to absorb and hold large amounts of water.

By the middle of the Triassic, about 230 million years ago, the sandstones were replaced by mudstones containing minerals, predominantly rock salt, which formed as deposits in shallow lagoon areas that were systematically evaporated and replenished from a marine source. Such rocks are called evaporites and they include rock salt (sodium chloride: NaCl) and gypsum (calcium sulphate: $CaSO_4$). Many of Europe's salt and gypsum deposits are from this period, for example the salt deposits in Cheshire and Germany. In Ireland the deposits include the salt at Carrickfergus in County Antrim. As well as its uses in cooking and road gritting, salt is a major raw material in the chemical industry, and gypsum is the main component of the plasterboard that covers virtually every wall in the interior of a modern building.

By the late Triassic there was a major re-advance of the sea, producing clay and calcareous deposits at the onset of the marine conditions that prevailed for much of the succeeding Jurassic period (210–145 million years ago). The Triassic witnessed the rise of the reptiles as the dominant life form on Earth, and footprints of an early reptile have been found in the Triassic sandstones at Scrabo quarries in County Down. These early reptiles evolved into the dinosaurs during the Jurassic, with the development also of the first true mammals.

The Jurassic

By the end of the Triassic (about 200 million years ago), Pangaea had begun to break up, with rifting taking place, producing flood basalts on continental margins. The first landmasses to break away were North America and Africa–South America (see Fig. 5.6e). This continental break-up was the precursor of a new ocean, the Atlantic, and the present distribution of continents stems from the break-up of Pangaea, which began at this time.

In Ireland the Jurassic is marked by the onset of shallow marine conditions rich in animal life, forming mudstone beds with some thin limestones. These deposits are best seen along the Antrim coast road; they contain bivalves, ammonites and occasional marine reptile remains. It is these soft mud layers (the Lias Clay) that are responsible for the unstable conditions along the Antrim coast road, which result in frequent landslides and road closures in that area. Throughout the Jurassic the fragmentation of Pangaea continued, affecting both the southern part (Gondwanaland) and the northern part (formerly Laurentia). For Ireland, the rifting of Laurentia and the eventual separation of North America from Europe were the events that were to have the greatest geological impact on the region in the periods following the Jurassic. Much of the rock record of the remainder of the Jurassic in Ireland is missing, probably lost to erosion.

The Jurassic has become fixed in the public imagination as the Age of Dinosaurs, with both carnivorous and herbivorous forms dominating life on land. During this time the earliest birds took to the air and mammals continued their still mostly insignificant role in the evolution of life.

The Cretaceous

As the break-up of Pangaea continued at pace (see Fig. 5.6f), the Earth entered one of its marine transgression phases, with raised sea levels on a global scale. It is estimated that, because of the fragmentation of Pangaea, the global area of shallow seas up to 50 m deep had doubled from the start of the Jurassic to Cretaceous times. Throughout much of the Cretaceous period, Ireland, along with most of Europe as far east as the Caucasus Mountains in Russia, was submerged under a warm shallow sea in which white limestone or chalk was deposited ("Cretaceous" is from the Latin *creta*: "chalk"). This white limestone, which forms the White Cliffs of Dover, is best exposed in Ireland along the Antrim coast road, where it forms spectacular cliffs, underlying the contrasting black basalt flows. The rock is composed almost entirely of the remains of shells of microscopic planktonic life forms called coccoliths. After death, the small calcareous discs contained in the cells fell to the sea bed in countless numbers, forming mud that eventually compressed to form the fine-grain limestone. As well as these microfossils, the Antrim chalk also yields sea urchins and cigar-shape fossils of squid-like animals known as belemnites.

Another obvious feature of the chalk deposits is the occurrence in them of well defined beds of **flint**. Flint is a form of silica, similar in many ways to quartz, and it originates from the silica needles or spicules that supported the soft forms of marine sponges. It is hard and durable, and, when broken, fractures in a characteristic way to give a very sharp edge. It is this property that was utilized by the earliest human inhabitants of the area to manufacture flint artefacts such as axes, arrowheads and scrapers.

The presence of flint nodules in such abundance was a major factor in the settlement of northeast Ireland, some 9000 years ago, by the earliest humans, and this is discussed on pp. 69–72. However, the final page of the Cretaceous chapter of the geological history of Ireland revealed a further mystery, which recent research is only now beginning to solve. Globally, the end of the Cretaceous was marked by yet another mass extinction of species, most notably in the public perception of that of the dinosaurs, but also the extinction of many other life forms, including in the oceans the tremendously successful ammonites. There is now substantial evidence that an impact involving a comet or asteroid colliding with the Earth may have been a major cause of this extinction. A huge impact crater has been identified at Chicxulub on the Yucatan Peninsula in Mexico. The effect of such a large-scale impact would be to discharge enormous amounts of gas and pulverized rock into the atmosphere, reducing the penetration of sunlight, thus causing catastrophic climate change, suppressing photosynthesis and thereby radically upsetting the food chain. That the effect was worldwide is shown by the identification of anomalously high values of the element iridium in clay layers at the top of the Cretaceous. This anomaly is found in many localities around the world, and the significance of iridium is that it is relatively rare in the Earth's crust but abundant in asteroids. Although firm evidence exists for such an asteroid impact, there may have been other factors involved. Associated with the splitting of Pangaea, there were major volcanic eruptions taking place globally, and the eruption of the Deccan Traps in India poured out enormous volumes of basaltic lava very quickly. A side effect of this would be the addition of vast quantities of gases to the atmosphere, with catastrophic consequences for the Earth's climate, similar to those produced by an asteroid impact. Whatever the reason or combination of reasons, this extinction marked the end of the Cretaceous and the Mesozoic era, and the beginning of the Cainozoic era.

The Palaeogene–Neogene

By about 70 million years ago, world sea levels had dropped again and northeast Ireland had reverted to a land surface. By 65 million years ago, at the end of the Cretaceous and beginning of the Palaeogene, Antrim was a low undulating limestone landscape with solution hollows and pockets of vegetation. Following the fragmentation of Pangaea and the subsequent break-up of the former Laurentia and Gondwanaland portions, the continents were beginning to resemble the shapes and distribution we recognize today (see Fig. 5.6g). By now, southern Africa was separated from South America by the southern Atlantic Ocean. India was a subcontinental island en route to collision with southern Asia to form the Himalayas, and the impending separation of North America, Greenland and Europe by the northward extension of the Atlantic Ocean was to dominate the geology of Britain and Ireland during the Palaeogene period.* In northeast Ireland, and also western Scotland, it is characterized by a prolonged period of igneous activity, particularly basalt lava eruptions, caused by greatly accelerated continental rifting as the North Atlantic developed and expanded, and North America separated from Europe and Greenland. This activity continues today along the line of the mid-Atlantic ridge, most noticeably in the very active volcanoes on Iceland.

At the beginning of the Palaeogene, Ireland was approaching the latitude of the present-day south of France (see Fig. 5.2). The break-up of the northern supercontinent Laurentia was proceeding, with the widening of the North Atlantic creating rifts between Europe and the east coast of Greenland, and between North America and the west coast of Greenland. Crustal thinning, which results in the eruption of large volumes of basalt lavas, precedes continental rifting. The remnants of the eruptions associated with the widening of the North Atlantic can be seen in Greenland, the

---

* Stratigraphical columns generally show the Cretaceous Period succeeded by the Tertiary Period . Some recent maps and publications, including those of the Geological Survey of Northern Ireland, have replaced the Tertiary with the Palaeogene and Neogene Periods. See Figure 4.3.

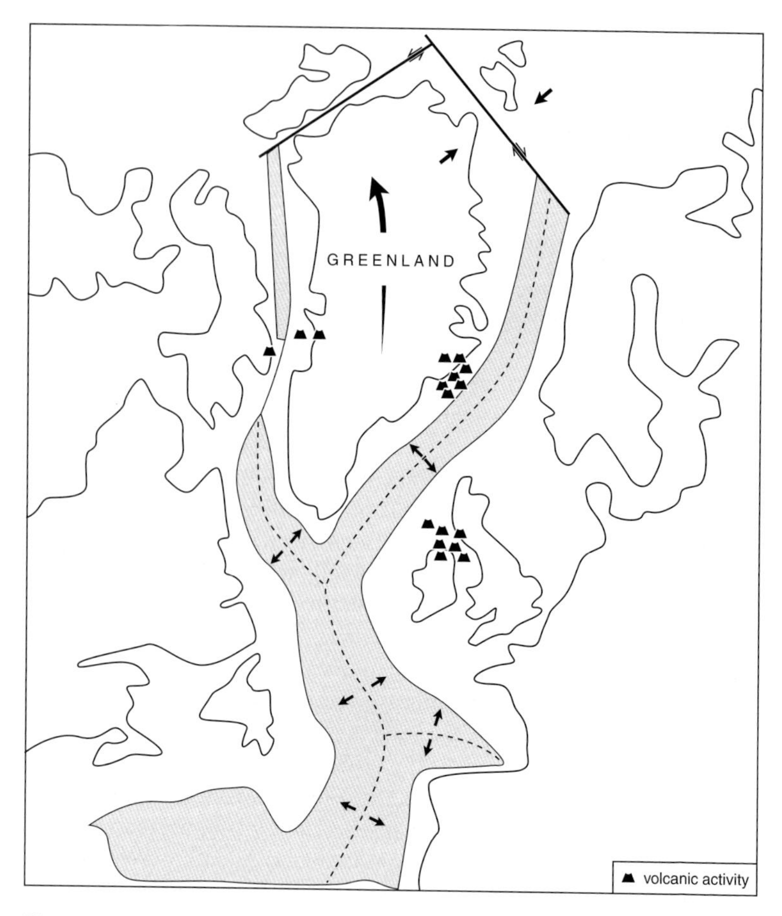

Figure 5.12 The extension of the North Atlantic in early Palaeogene times (after Harland 1969).

Hebridean islands of Mull and Skye and, most notably for this account, the basalt lavas of the Antrim Plateau and the Causeway Coast (Fig. 5.12).

The eruptions began in northeast Ireland with a phase of explosive volcanism. This left thick deposits of volcanic ash and agglomerate, which are conspicuous in north Antrim on the island of Carrickarede, perhaps better known for its tenuous connection to the mainland by a precarious rope bridge. After this, prolonged eruption of basaltic lavas from elongate fissures or cracks produced the Antrim Lava Group, forming the Antrim Plateau. Pauses in eruption allowed surfaces within the succession to become deeply weathered, producing bright red layers of oxidized basalt known as **laterites**, one of which is spectacularly exposed at the Giant's Causeway. The Antrim lavas include the characteristically columnar forms to be found both at the Causeway and on the Causeway Coast.

While this eruptive phase continued in Antrim, igneous activity took place some 100 km farther south in counties Down, Armagh and Louth. Unlike the fissure eruptions that produced the Antrim lavas, the volcanoes of Slieve Gullion and Carlingford were central volcanoes. As the term implies, these are roughly circular in shape and probably formed a conical volcano with a central vent. The nearby granite Mourne Mountains show geochemical and geophysical evidence of their being derived from a gabbro intrusion deep beneath south County Down, and a total of five chemically distinct magma types can be identified in the Mournes. However, there is no evidence that any of these magmas ever reached the surface as lavas, remaining instead as intrusions that cooled in the upper levels of the crust and are exposed now as a result of over 60 million years of erosion.

The final stage of these cataclysmic volcanic and plutonic events, which so influenced the landscape of northeast Ireland, was the intrusion of more relatively small-scale sills and dykes right across the northern half of Ireland. After this, igneous activity in northeast Ireland waned and the mantle plume that produced it is now marked by the concentration of active volcanoes in Iceland on the mid-Atlantic ridge.

After the turbulent early Palaeogene, later Palaeogene times were relatively quiet. As is common with many lava-plateau areas, crustal subsidence occurred after the eruptive phase, forming a generally saucer-shape basin in northeast Ireland, with Lough Neagh, the largest body of fresh water in Britain and Ireland, at its approximate centre. Erosion of the higher ground around this basin produced much of detrital material that accumulated as the Lough Neagh Clays. These rocks are the youngest bedrock in northeast Ireland. Up to 300 m thick in places, they are potentially important as a source of fuel, as they contain considerable reserves of lignite (brown coal), which formed by the accumulation of vegetation in the Lough Neagh basin. The absence of sediments younger than Palaeogene age suggests that the succeeding Neogene was a period of erosion across the north of Ireland and no rocks from this time survive today. The Neogene is succeeded by the **Quaternary period** (1.6 million years ago to the present day), the first subdivision of which is the **Pleistocene** (1.6–0.01 million years ago).

The Quaternary

From an Irish perspective the most significant Pleistocene geological event was the formation of large mid-latitude ice sheets. The erosion and deposition effects of this ice age were to have a dominant influence on the

Quaternary landscape. Many of the landforms we see today are the products of this phase of geomorphological development. Much of the scenery of the northern part of Ireland is related to the final phase of this glaciation, which lasted from about 25 000 to 17 000 years ago.

Glacial erosion characteristically produces valleys with a broad U-shape cross section. These can be seen in localities such as Glenariff in County Antrim and Glenveagh in County Donegal. The predominant superficial deposit over the whole of the north of Ireland is **boulder clay**, which formed below the ice sheet and consists of a stiff clay containing boulders of all sizes. Over much of the area of counties Down, Armagh, Monaghan and Cavan, the clay deposits were sculpted by the ice into small rounded hills called **drumlins**. During periods of ice melting, rivers flowing out from the icecaps left deposits of sand, silt and clay. These often formed delta-like features such as those seen around Ballyvoy, east of Ballycastle, or they formed the sinuous ridges of sand and gravel known as **eskers**, found extensively throughout Ireland, with many good examples in Tyrone and Fermanagh. As the ice sheets disappeared, the area experienced a tundra landscape and climate similar to that of present-day Arctic Canada or Russia. Finally, about 10 000 years ago, a general global warming heralded the beginning of the current warm stage, the **Holocene** or Recent, which has been relatively stable climatically.

The end of the Ice Age was marked by fluctuations in sea level as icecaps melted and the land rose as the vast weight of ice diminished. Over thousands of years this weight had depressed the crust and, following the melting of the ice, it began to rise gradually. However, because the icecaps melted rapidly over a relatively short time, the water level rose faster than the crust. This meant that for a short period the postglacial sea levels rose and were much higher than at present, possibly up to 100 m higher in places. Raised beaches are evidence of these higher sea levels; they are characterized by such features as cliffs, stacks, caves and arches at various levels above the present high-water mark, for example, west of Ballintoy Harbour, as described in the North Antrim excursion (pp. 161–168).

The progressive improvement in climate over the past 10 000 years enabled widespread tree coverage to develop over Ireland, reaching a peak about 7000 years ago, with a wide range of plant species and a varied fauna, including deer, boar, wolf and fox. Following this climax in forest growth, a change to a wetter climate favoured the development of peat bogs over much of the region, a change partly influenced by forest clearance for agriculture, undertaken by the earliest inhabitants of the island. This was the beginning of the process of landscape modification by human activities, a force increasingly important in shaping the world around us.

In plate-tectonic terms, seafloor spreading continues and Ireland is still drifting generally northwards. Currently, we are remote from plate edges and therefore free from volcanic and earthquake activity, but in geological terms this is merely a brief interlude before the next phase of continental rifting or collision brings Ireland back into the forefront of plate tectonics.

Chapter 6

Geological maps

Just as topographical maps are used to help navigate around the Earth's surface, so geological maps are used to explore and clarify the surface and subsurface distribution of rocks. It would be helpful at this stage to consider some of the particular features of geological maps that make them different from normal topographical maps, with which the reader may be more familiar.

Making geological maps

The most common aim of geological fieldwork is to produce a geological history of the area under investigation by the geologist, and the summary of that history is the geological map. Like the topographical maps on which they are based, geological maps can be presented at a range of scales, but they all share certain features. Some degree of familiarity with these common features will allow the reader to interpret the geology of the area at several different levels. The geological map shows the distribution on the surface of the solid bedrock (**solid map**) or of the superficial deposits such as peat, boulder clay and sand (**drift map**). This guidebook refers mostly to the 1:250 000 scale solid-geology map of Northern Ireland, produced by the Geological Survey of Northern Ireland (GSNI).

At its simplest level, a geological map can be used to identify the rock type occurring on the surface at any point on the map. However, with some experience and using additional information obtained from the map, such as the sequence in which the rock types occur, it is possible to interpret the three-dimensional structure of the upper part of the crust. It may also be possible to predict the subsurface geology underneath any covering of superficial deposits, vegetation or buildings. Because the rocks are laid down in sequence, the map also allows one to look back through time and study the evolution of the landscape. In northeast Ireland, former geological environments, including hot deserts, shallow coral seas, deep oceans and active volcanoes, can be recognized in the rock successions and

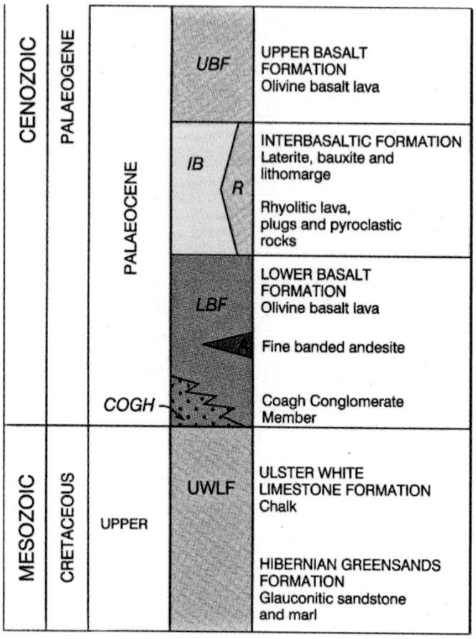

CENOZOIC	PALAEOGENE	PALAEOCENE	*UBF*	UPPER BASALT FORMATION Olivine basalt lava
			IB / *R*	INTERBASALTIC FORMATION Laterite, bauxite and lithomarge Rhyolitic lava, plugs and pyroclastic rocks
			LBF	LOWER BASALT FORMATION Olivine basalt lava Fine banded andesite
			COGH	Coagh Conglomerate Member
MESOZOIC	CRETACEOUS	UPPER	UWLF	ULSTER WHITE LIMESTONE FORMATION Chalk HIBERNIAN GREENSANDS FORMATION Glauconitic sandstone and marl

Figure 6.1 The stratigraphy of County Antrim. This extract and all subsequent map extracts are from the *Geological map of Northern Ireland*, 1:250 000 Series (2nd edn, 1997), published by the Geological Survey of Northern Ireland and reprinted with permission (© Crown copyright).

are shown on the appropriate geological maps. For example, the sequence of rocks in County Antrim in Figure 6.1 shows the Ulster White Limestone (Chalk) overlain by the various basalt and laterite formations of the Antrim lavas, illustrating the change from the warm shallow seas of Cretaceous times to the volcanic eruptions that built the Antrim Plateau in the succeeding Palaeogene times.

Most geological maps are based on observations made on the ground by a geologist who identifies and records the outcrops of bare rock. These may be natural occurrences or manmade features such as road cuttings or quarries. Information from boreholes and wells may also be used. Between areas of good exposure, boundary lines are drawn by inference and interpretation, using features such as changes in slope or landform, or variations in vegetation cover. Mapping at this stage is often at a scale of 1:10 000 (10 cm to 1 km) and the information is then used to produce maps on the 1:50 000 scale (2 cm to 1 km) and other scales such as the 1:250 000 scale map (2 cm to 5 km) this guide refers to. As well as the bedrock mapping, the geologist will normally record the occurrence of the superficial deposits such as peat or boulder clay, referred to as drift.

Reading geological maps

The geological column for Ireland shows that geological time is subdivided into units (Fig. 4.1). The most commonly used unit is the period, which is a worldwide unit characterized in some particular way. For example, the Devonian period occurred from about 410 million years ago to 350 million years ago and consists typically of desert-derived sediments that formed in a hot desert environment. Several periods may be grouped together to form an era; for example, the Devonian is one of six periods in the Palaeozoic era, succeeded by the Mesozoic and Cainozoic eras. The largest division of geological time is the eon; for example, the Phanerozoic, from 600 million years ago to the present, comprises the Palaeozoic, Mesozoic and Cainozoic eras.

To aid the description of rocks and their representation on the map, geologists use a system of rock units referred to as **members, formations** and **groups**. The fundamental unit of this hierarchy is the formation, which consists of a sequence of related rock types that can be mapped together and which may be composed of smaller units called members. Sequences of related formations may be combined into a group, and groups into **supergroups**. These units in sequence are referred to as the stratigraphy of the area and are represented by the stratigraphical column shown on the map. This can be illustrated by reference to the Slieve League area in Donegal, to be featured in one of the field excursions (see pp. 87–95). Figure 6.2 shows the stratigraphy of the Dalradian rocks of Donegal and Sligo. The position of a particular unit in the succession is shown with a specific colour and letter code indicating the rock type; by convention, the oldest rocks are at the bottom of the column and become progressively younger upwards. In the Slieve League area the rocks examined in the South Donegal excursion (pp. 87–95) include the following formations: Slieve Tooey Quartzite Formation, Port Askaig Formation, Glencolumbkille Formation

Figure 6.2 shows that the Slieve Tooey Quartzite Formation and the Port Askaig Tillite Formation are part of the Argyll Group, whereas the Glencolumbkille Limestone Formation is part of the Blair Atholl Group. Both these groups are part of the **Dalradian Supergroup**. It can be shown in the field that the Glencolumbkille Limestone Formation can be further subdivided into two members: the Glenhead Schist Member and the Glencolumbkille Dolomite.

Era		GROUP	SUBGROUP	Ox Mountain County Sligo		Inishowen/Slieve League North County Donegal		South and Central County Donegal	
PROTEROZOIC	DALRADIAN	SOUTHERN HIGHLAND GROUP				GREENCASTLE GREEN BEDS	IB	SH	UPPER DALRADIAN (Undivided)
						INISHOWEN HEAD GRITS	IG		
						CLOGHAN GREEN BEDS	CI	CG	CROAGHGARROW FM ≡ Deele Group in Central Donegal
						FAHAN GRITS	FG	SG	SHANAGHY GREEN BED FORMATION
						FAHAN SLATES	FS	MY	MULLYFA FORMATION
		ARGYLL GROUP	TAYVALLICH			CULDAFF LIMESTONE	CL	AY	AGHYARAN FORMATION
			CRINAN			UPPER CRANA QUARTZITE	UC	KT	KILLETER QUARTZITE
						LOWER CRANA QUARTZITE	LC		
			EASDALE	UMMOON FORMATION	UM	TERMON PELITE	TM	LE / LM	TERMON PELITE LOUGH ESK PSAMMITE LOUGH MOURNE GRIT
				CARRICK O'HARA FM	CO	CRANFORD LST	SC	BG	BOULTYPATRICK GRIT
						SLIEVE LEAGUE FM	SL	CO	CROAGHUBBRID PELITE
								RF	REELAN FORMATION
			ISLAY			SLIEVE TOOEY QUARTZITE	ST	GA	GAUGIN QUARTZITE
						PORTASKAIG TILLITE	BB	BB	
		APPIN GROUP	BLAIR ATHOLL			GLENCOLUMBKILLE LST	QC	OC	OWENGARVE CALCAREOUS FM
						FINTOWN PELITE	FI		
						GLENCOLUMBKILLE PELITE FORMATION	QP		
						LOUGHROS GP AND Ur FALCARRAGH PELITE	LF		
						FALCARRAGH LIMESTONE	FL		
						Lr FALCARRAGH PELITE AND CLONMASS FM	FC		
			BALLACHULISH			ARDS QUARTZITE	AQ		
						ARDS BLACK SCHIST AND CREESLOUGH FM	AC		

Figure 6.2 Stratigraphy of the Dalradian rocks of County Donegal (© Crown copyright).

The 1:250000-scale geological map

The *Solid geology of Northern Ireland* map is complex, covering a large area at a scale of 5km to 2cm (about 4 miles to the inch). However, using the key map and the related stratigraphical column, it can be subdivided into six key stages of the stratigraphical column. These stages can be related to specific geographical areas, so the map shows the distribution, relative age and stratigraphy of the principal geological divisions over the whole region (Fig. 6.3); the numbers below refer to the numbers on the key map.

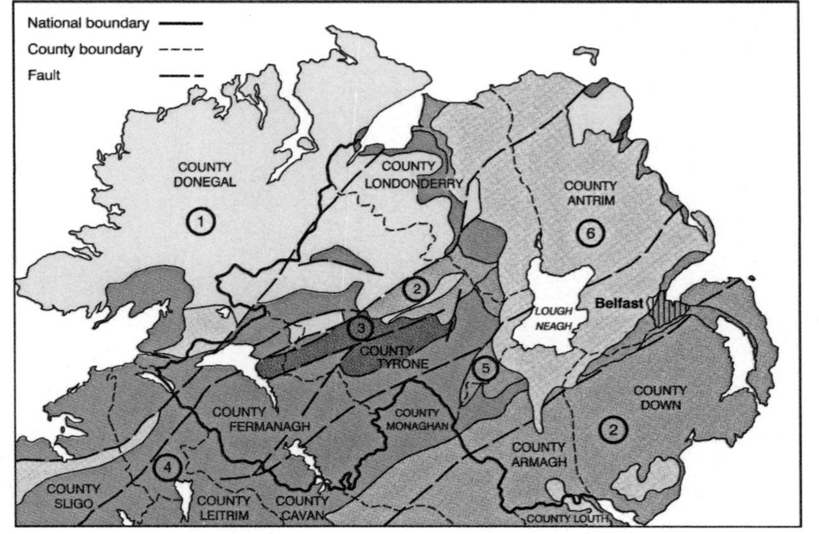

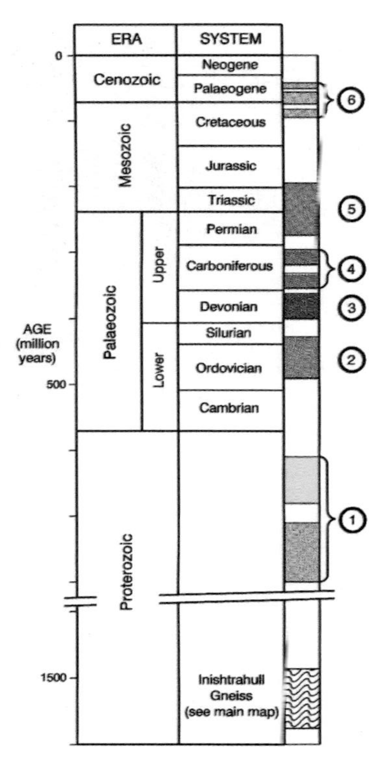

Figure 6.3 Key map showing distribution, relative age and stratigraphy of the principal geological divisions of the north of Ireland. The numbers on the map and the stratigraphical column also relate to the detailed stratigraphy presented in the index to colours. (© Crown copyright.)

1. **Proterozoic** rocks are the oldest in the region; mainly **Dalradian**, they are found principally in the Donegal–Sperrins area to the west, with a smaller area around Torr Head in north Antrim. They are predominantly metamorphic rocks, mostly schists, gneisses, quartzites and marbles.
2. **Ordovician–Silurian** sediments, mainly sandstones and mudstones forming the bedrock of most of counties Down, Armagh, Monaghan and Cavan.
3. **Devonian** rocks, occurring mainly in the Clogher Valley in Tyrone and around Cushendall–Cushendun in east Antrim. They comprise mainly desert-derived sediments.
4. **Carboniferous** rocks, occurring most extensively in counties Fermanagh, Sligo and South Donegal, but with important occurrences also in counties Leitrim, Cavan, Monaghan and Tyrone. Comprising mostly limestones, rocks of Carboniferous age are the most common bedrock in Ireland.
5. **Permo-Triassic** rocks, occurring principally in north Down and the Lagan Valley around Belfast, where the sandstones form an important aquifer or source of groundwater, but there are also important localities in counties Tyrone and Londonderry. Like the older Devonian rocks, these are predominantly desert-derived sediments.
6. **Cretaceous** and **Palaeogene** rocks, occurring in the eastern half of the region. (Rocks of Palaeogene age may be referred to on other geological maps as Tertiary. The Tertiary period succeeds the Cretaceous and has been subdivided into an older part, the Palaeogene, and a younger part, the Neogene). The Cretaceous rocks are the white limestones of the Antrim coast area, where they underlie the Palaeogene basalt lavas of the Antrim Plateau. Other Palaeogene igneous rocks are the Slieve Gullion volcano in South Armagh, the Carlingford volcano in County Louth, and the granites of the Mournes in south Down.

The key map also shows some of the main fault lines such as the Tow Valley Fault and the Omagh Fault, which affect the rocks in this area and have probably been major structural influences on the landscape throughout geological time.

Geological boundaries

To someone attempting to read a geological map for the first time, the proliferation of colours and lines can seem daunting. However, by using the key map and the stratigraphical column to identify the main ages and lithologies involved, and by recognizing that different types of boundary

Era	Period	Epoch	Unit	Description
MESOZOIC	JURASSIC	LOWER	WMF	WATERLOO MUDSTONE FORMATION Grey mudstone and thin limestone
	TRIASSIC	UPPER	PNG	PENARTH GROUP Dark grey mudstone
		UPPER, MIDDLE	MMG	MERCIA MUDSTONE GROUP Red-brown and green mudstone and marl with thick salt beds in South Antrim
		LOWER	SSG	SHERWOOD SANDSTONE GROUP Red-brown sandstone
UPPER PALAEOZOIC	PERMIAN	UPPER	BELF	BELFAST GROUP Marl with gypsum, dolomitic limestone, basal sandstone
		LOWER	ENLE	ENLER GROUP Red-brown sandstone, conglomerate and siltstone, basal breccia

Figure 6.4 Conformable series of Triassic rocks overlain by the Lower Jurassic Waterloo Mudstone Formation (© Crown copyright).

exist between rock types, it is generally possible to break the map up into smaller units as an aid to understanding the geology.

For the beginner the most important contacts to recognize are the stratigraphical boundaries between the various sedimentary rock units making up the map. **Conformable** sequences are those sedimentary sequences that follow on from one another without any appreciable time break in the sequence. For example, the Sherwood Sandstone Group, the Mercia Mudstone Group and the Penarth Group of Triassic rocks that are found in the Lagan Valley area around Belfast form a mainly conformable sequence of sediments, succeeded **unconformably** by the Jurassic Waterloo Mudstone Formation (Fig. 6.4).

Unconformable boundaries are those that mark a break in the rock sequence, often an erosion period of considerable duration, as illustrated by the case study at Murlough Bay, discussed on pp. 25–29 (see Figs 4.6, 4.7). Another good example of an unconformity is the small outcrop of Devonian rocks at Ballymastocker Bay on the Fanad Peninsula in Donegal, which lies on Dalradian rocks that are about 130 million years older, and during that time the Dalradian rocks were folded and metamorphosed (Fig. 6.5).

Faulted contacts are the lines of fracture that have broken through sequences of rocks and moved them horizontally or vertically. These can be on the scale of a few metres to tens or hundreds of kilometres. The Tow Valley Fault (Fig. 6.6), which is clearly seen on the map trending southwest

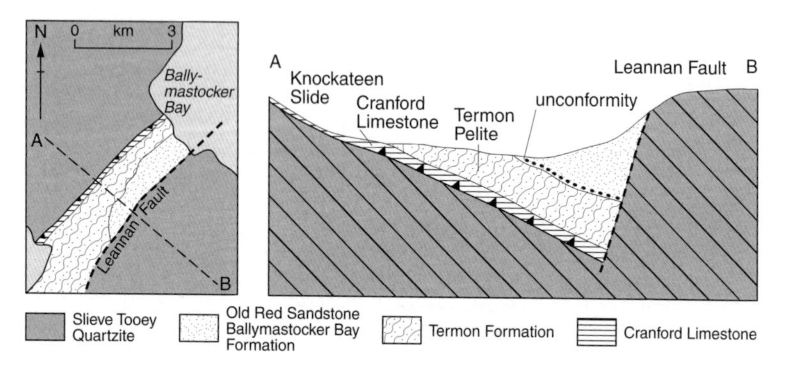

Figure 6.5 The Dalradian/Devonian unconformity at Ballymastocker Bay, Fanad, County Donegal.

from Ballycastle on the north Antrim coast, separating Dalradian metamorphic rocks from younger Palaeogene volcanic rocks, is probably a continuation of the Highland Boundary Fault in Scotland to the northeast. This major fault line is thought to continue southwestwards to the west coast of Ireland at Clew Bay in County Mayo and it probably represents a terrane boundary formed during the continental collisions that closed the Iapetus Ocean in Silurian times (see Fig. 5.9). Examples of boundaries formed by igneous bodies are illustrated below. Intrusive igneous bodies often cut across pre-existing stratigraphical boundaries; they are discordant intrusions (Fig. 6.7). Sheet-like horizontal igneous bodies such as sills may have intruded along the stratigraphical boundaries and lie parallel to them; they are concordant intrusions (Fig. 6.8). In contrast to the sill, the nearby Palaeogene dykes cut across the sedimentary boundaries in the area and are therefore discordant.

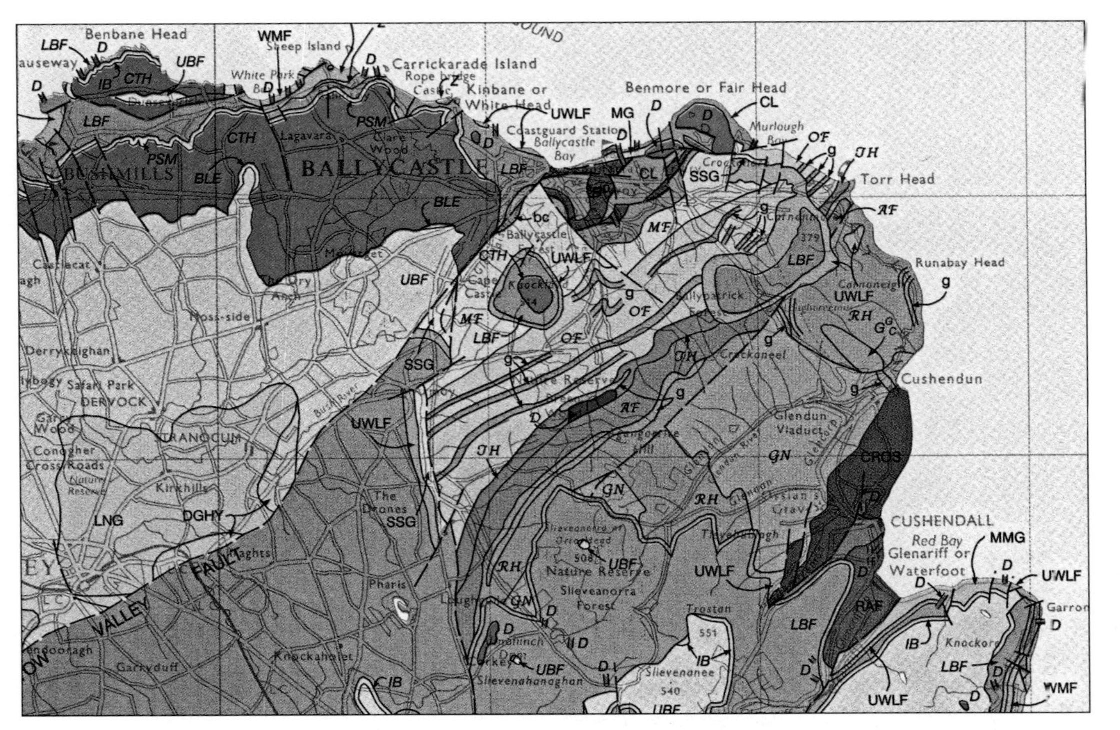

Figure 6.6 Tow Valley Fault separating Dalradian rocks from much younger Palaeogene volcanic rocks to the north (© Crown copyright).

Figure 6.7 The Newry Granite intrusion – a discordant intrusion. Here the outline of the Newry intrusion clearly cuts across several boundaries of the Ordovician and Silurian sediments making up the country rock intruded by the granite. (© Crown copyright.)

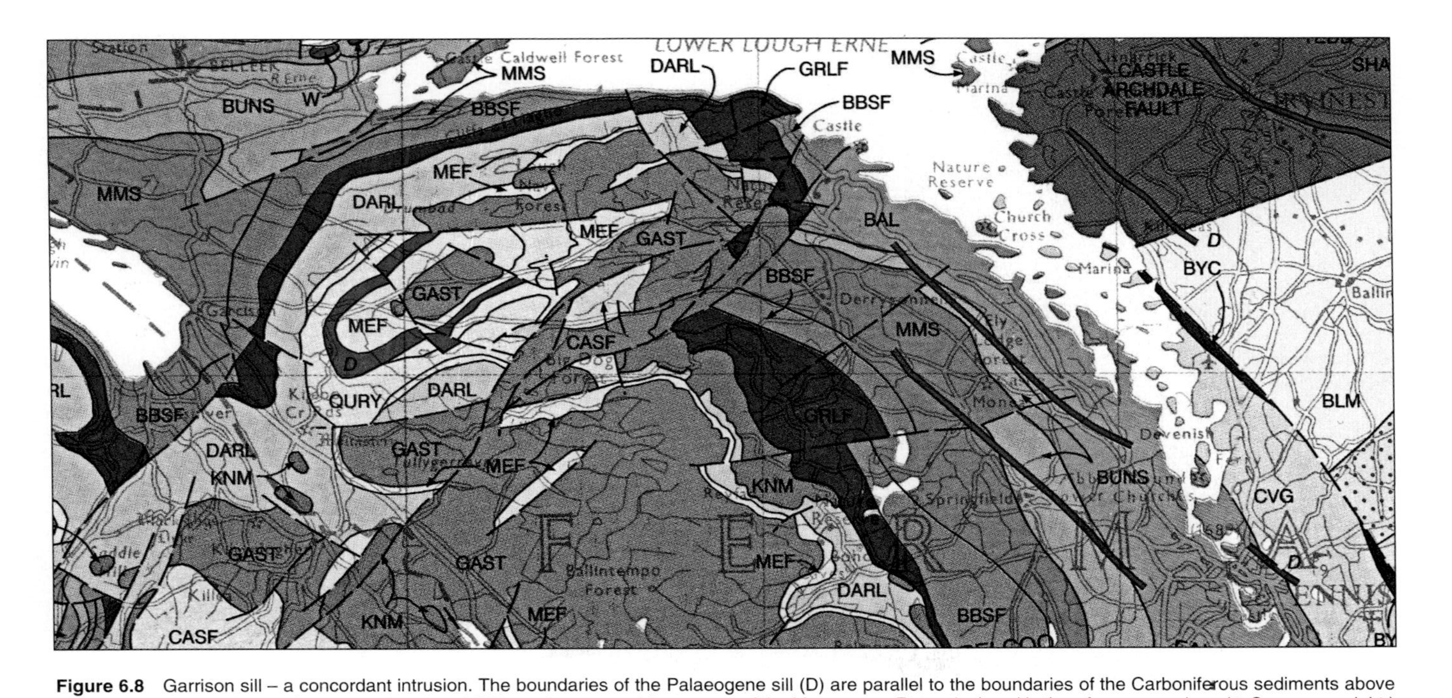

Figure 6.8 Garrison sill – a concordant intrusion. The boundaries of the Palaeogene sill (D) are parallel to the boundaries of the Carboniferous sediments above and below it (QURY: Quarry Sandstone Member; MEF: sandstones and limestones of the Meenymore Formation) and is therefore concordant. (© Crown copyright.)

Chapter 7

Geology and archaeology in Ireland

Although this book is a guide to the geology of the north of Ireland, it is interesting to examine the links between the earliest settlers in Ireland and the rocks making up their landscape. The science of archaeology has revealed much about lifestyles during the gradual colonization of Ireland since about 9000 years ago, and there is often a geological dimension to many aspects of the activities of these early societies, which is frequently ignored or undervalued. Early human beings demonstrated an awareness of the geology of the inhabited area as a source of raw materials such as flint and **chert**, useful for the manufacture of tools and weapons. There was also an appreciation of the various component parts of the landscape. This is made obvious by the addition of built structures to prominent parts of the landscape from earliest times, thereby emphasizing and embellishing those landscape features, a good example being a large cairn (Maeve's Tomb) on the prominent hill Knocknarea, near Sligo. This feature is clearly designed to be visible from particular directions, including the major Mesolithic cemetery at Carrowmore, about 3 km to the west (Fig. 7.1).

The recognition of the main elements of the landscape, the selection of large blocks of stone for the building of monuments, the seeking out and

Figure 7.1 Maeve's Tomb on Knocknarea, viewed from the Megalithic cemetery at Carrowmore, County Sligo.

exploitation of specialized rock resources such as metal ores, and the use of rocks for mundane uses such as the grinding of grain – all require the skills and techniques of today's geologist. Modern society still depends on the exploitation of the rocks of the region for building materials, road-stone, drainage pipes and a wide range of other requirements.

In the evolutionary progression from hominid forms to the appearance of *Homo sapiens*, which has taken place over the past few million years, one of the features that has characterized the higher forms of human life has been the ability to manufacture and use tools, many of which, from earliest occurrences, were pieces of rock. The use of the term "Stone Age" is an indication of the importance of rock, and therefore geology, in the development of early man. It could be said, therefore, that the human species had a very early interest and expertise in matters geological.

From primitive beginnings, the use of rocks as tools and cultural artefacts developed to a remarkable degree of sophistication. This can be seen in the finely worked flint arrowheads and javelin heads that have been recovered from some of the early burial chambers found in many parts of Ireland. Examples of the role of rocks and minerals in early social development in Ireland are widespread, from the use of flint nodules to make cutting and scraping tools, to the building of great stone monuments such as those at Carrowmore in Sligo. Here it appears that the builders were aware of differences in rock types and were able to differentiate metamorphic rocks, such as gneisses, from igneous rocks, such as granite, and sediments, such as limestone. This chapter examines the uses made of various rock types during the social and cultural evolution of the north of Ireland.

The Stone Age: the Mesolithic period

The first settlers in Ireland probably arrived on the island about 9000 years ago, in the Mesolithic or Middle Stone Age (Fig. 7.2). Evidence of any earlier settlement has not been found and may have been completely obscured by the processes operating at the end of the Ice Age. It is also likely that Ireland then was already an island, cut off from Britain and the European continent by the rise in sea levels at the end of the glaciation. The early inhabitants were hunter–gatherers who had a wide range of food-stuffs to sustain them, including fish in the rivers, game such as deer and pig, and a variety of edible plants and shellfish. The early environment seemed to be fairly densely wooded, and open water or grassland areas were relatively limited. Mesolithic people therefore tended to choose riverbanks and estuaries or lake margins as settlement sites. One of the oldest sites investigated is at Mount Sandel on the banks of the River Bann, near Coleraine. With calculated dates around 9000–8650 years ago, this is the earliest known occupation site in Ireland. Flint **microliths** and other flint artefacts have been recovered from the Mount Sandel excavations. Microliths are very small artefacts, normally of flint or chert, mounted in grooves in wooden hafts, or occasionally as points. The significance of the Mount Sandel finds is that from the earliest known settlement site in Ireland there is already evidence for a sophisticated toolmaking culture using the flint nodules found in the Cretaceous chalk deposits that underlie the basalts in much of northeast Ireland. Many of the earliest settlement sites in Ireland were in Antrim and Down, which may be attributable to the close proximity of Scotland, but may also have been influenced by the availability of flint from the nearby Cretaceous chalk deposits.

The Neolithic period

The development of agriculture, probably around 6000 years ago, brought about a new way of life in Ireland, with a change from a principally hunter–gatherer mode to one in which there was a more settled life style (Fig. 7.2). Surplus supplies were stored to bridge the leaner winter months, and more complex tools and implements were developed. The manufacture of specialist tools such as polished stone axes, and the building of vast stone tombs that must have required a high level of organization to design and build, were further indications of an increasingly sophisticated society. The production of polished axe heads not only reflected the need for more efficient means of clearing forest, as the need for farming land increased, but also involved ceremonial uses for the axes, and also the first signs of trade outside the immediate area of Ireland. All of these developments imply that a complex social order was in place by this time.

The development of agriculture meant a need for a wider range of tools

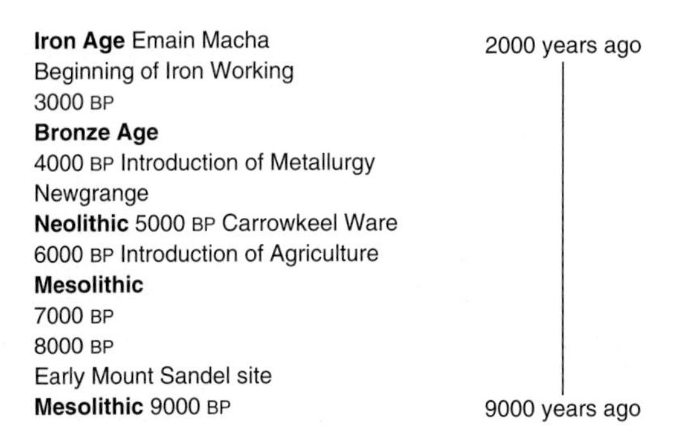

Figure 7.2 Time line for the north of Ireland from early Mesolithic to the Iron Age.

and implements, and this required recognition of various rock types and their ranges of properties. For example, grain requires some method of grinding to convert it into an edible form. This was done using a quern: two pieces of stone used as grinders. In the simplest, the saddle quern, the lower stone has a concave top in which the grain is placed before being rubbed by the upper rock. The rotary quern operated by causing a circular upper stone to rotate on a fixed lower one. By dint of the job they are designed to do, querns require coarse abrasive material, and a commonly used rock was a coarse-grain, white, quartz-rich sandstone found in the transition beds between the Devonian and Carboniferous periods.

Relatively soft rocks such as schist, and slate or shale, which tend to break into flat plate-like fragments, were used when the requirement was for perforated line sinkers or loom weights. As mentioned earlier, flint occurs as nodules along the bedding planes of the Cretaceous chalk in northeast Ireland. In addition, it occurs in glacially derived sediments and in beach deposits as a result of Scottish ice moving down the Irish Sea basin and gouging out flint-bearing rock from the sea floor and depositing it along with boulder clay on the east coast of Ireland. Chert is another form of silica and it is physically and chemically indistinguishable from flint. Whereas flint occurs in Cretaceous limestone, chert is found in limestones of Carboniferous age. Most of the artefacts recovered from the megalithic monuments around Sligo were therefore manufactured from chert.

Perhaps one of the most impressive examples of the geological expertise of the Neolithic settlers in Ireland around 5000–6000 years ago was their use of a rock named **porcellanite** to produce very efficient polished stone axes. This porcellanite occurs in only two places in northeast Ireland: at Tievebulliagh, a Palaeogene plug west of Cushendall in Antrim, and at the Brockley Plug on Rathlin Island. At both these localities the porcellanite (so named because its fine-grain blue appearance resembles porcelain) was formed by thermal metamorphism or heating of weathered basalt (laterite) in an adjacent volcanic vent. As both these localities show relatively small amounts of porcellanite and are in obscure and remote localities, it is a measure of the observational powers of Neolithic man that they were found and exploited. The hard brittle porcellanite was probably extracted by first using fires to heat the rock, then chilling with water to cause fracturing, thus allowing workable blocks to be extracted from the outcrop. These were shaped on site into the general form of an axe, possibly by using quartzite hammer stones. Final polishing, probably against hard sandstone blocks, was carried out at axe "factories" elsewhere. These axes were used extensively in Ireland and some of the larger examples may have had a purely symbolic or ceremonial function. There is also considerable evidence that many of these axes were exported to all parts of

Britain and Ireland. Five thousand years ago the production of porcellanite axes was an industrial undertaking, with the earliest settlers acting as exploration geologists, quarrymasters, tool manufacturers and exporters – precursors of today's geological consultants, quarry operatives and manufacturers and retailers of concrete products.

The Megalithic builders

Although the use by the early inhabitants of rock materials can be understood in purely practical terms, it is more difficult to appreciate fully the significance of their more spectacular use of rocks to build a variety of stone monuments. During Neolithic times a culture of building large tombs developed and, since the blocks of stone used to build these monuments are often huge, the general term to describe them is **megaliths**, from the Greek words *megas* ("large") and *lithos* ("stone"). They assume a variety of forms, named after their principal architectural features: court tombs, portal tombs, passage tombs and wedge tombs. The earliest and simplest is probably the portal tomb, in which upright stones are covered by a capstone that frequently weighs many tonnes. A simple passage tomb stands at the centre of Giant's Ring, near Belfast (Fig. 7.3).

The Neolithic cemetery complex at Carrowmore in County Sligo is the largest single cluster of passage tombs in Ireland, and modern archaeological interpretation now recognizes the relationship between the layout of the tombs and the hills of Knocknarea to the west and Carn Hill to the east. The spatial arrangement of the tombs was obviously critical and it

Figure 7.3 Passage tomb at the centre of the Giant's Ring, County Down. © Crown copyright; reproduced with the permission of the Controller of HMSO.

seems likely that there was ceremonial significance both at the level of individual tombs and also involving the larger-scale layout of the whole complex and the surrounding landscape.

The sheer scale of these megalithic monuments indicates a high level of engineering skill, with a corresponding high level of social organization to ensure the co-operation needed for such major engineering feats. It is also clear that the visibility of monuments was somehow important for perhaps religious, social or political reasons, and it indicates an early appreciation of not only the overall landscape but also the component rock types that make up that landscape.

It seems that the recognition of specific rock types and the various uses to which they could be put – functional, decorative or ceremonial – played an important role in the social, cultural and eventual industrial development of the people of Ireland from earliest times to the present.

The geology of the Bronze and Iron Ages

The Bronze Age succeeded the Neolithic Period in Ireland around 4000 years ago, the advent of metallurgy coinciding with some major changes in the environment. These changes included the development of thick peat deposits and the formation of blanket bogs that were to have such profound landscape implications subsequently. It appears that there was a trend towards wetter and cooler climatic conditions, with the spread of blanket bog mainly attributable to the impact of farming on vegetation and soils over a long period. As water tables rose and soils became wetter, plant remains such as *Sphagnum* mosses began to accumulate on the surface and eventually buried large expanses of the countryside.

Bronze production

Bronze is a mixture of copper and tin. Within Ireland, copper ore is limited to areas around Cork and Kerry, the Waterford coast, Avoca in county Wicklow and in Tipperary. The area under consideration here, the north of Ireland, is virtually devoid of copper, except for small amounts occurring here and there as "native copper", native metals being those that are already in the metal form and so do not need any processing or smelting. Native copper is now very rare in Ireland and was probably never a major source of copper in prehistoric times. In the areas in the south of Ireland, where the various copper ores occurred, a smelting technology was developed, at first using copper ores that were oxides and which occurred near the surface and were readily obtainable. When these ores were exhausted, more complex processes involving roasting the ore were developed to allow exploitation of the underlying sulphide ores. Later in the Bronze Age, tin was alloyed with the copper to produce true bronze implements (typically 10% tin, 90% copper). As tin ore (cassiterite) is very rare in Ireland, it is likely that the later Bronze Age artefacts were made using tin imported from Cornwall.

The distribution of copper mines in the southwest of Ireland suggests a detailed and systematic search for the ore, and indicates, as with the exploitation of porcellanite in the Neolithic, a sophisticated awareness of the rock formations containing the ore beds. Those involved in organizing the exploration, extraction and processing of the copper ore would have been an important group within Bronze Age society and they were also probably involved in the distribution and export of the copper metal, as well as the manufacture of implements from it. Many Bronze Age artefacts were made by casting and, as the artefacts became more complex, so the demands made on the moulds increased. For example, initially flat axes were cast in simple open moulds carved from sandstone or similar rock types. These evolved into two-piece moulds carrying a central clay core to produce weapons or tools with hollow sockets that allowed the metal head to be fitted to a wooden handle. Relatively soft rocks such as talc schist (talc is one of the softest minerals known) or steatite, a rock composed mainly of talc and sometimes called soapstone, were used to produce more elaborate moulds, perhaps with surface markings on the implements. In the later part of the Bronze Age there was a move towards the use of clay moulds, and recognition of suitable clay deposits for moulding or pottery was a further extension of the geological repertoire of the early settlers in Ireland.

Iron production

Unlike the relatively localized distribution of copper ores in Ireland, iron ores are common and they occur in a variety of forms and environments. In the north of Ireland they are found as ironstone nodules (iron carbonate) in the Carboniferous coal areas such as Ballycastle and Coalisland, in weathered beds of basalt known as laterites (iron hydroxides) in the Antrim basalts and as bog iron (hydrated iron oxide) in the widespread blanket bogs. This ubiquitous distribution meant that Iron Age metal workers could set up their smelters almost anywhere. A mixture of clay-ironstone nodules and bog iron ore, similar to that available in the Iron Age, was still being used in Irish iron production up to the eighteenth century.

Gold production

The metalworking skills developed to produce copper and bronze implements in the Bronze Age were also extended to the manufacture of decorative objects from gold. Gold occurs in several localities throughout Ireland, but many of today's known sources are in the bedrock and would not have been available to Bronze and Iron Age miners. Most of the gold utilized in the Bronze Age would have come from so-called "placer" deposits, that is, gold flakes or nuggets of gold in river sands or gravels and which is typically retrieved by gold panning. Even today it is possible to recover small amounts of gold from rivers in the Sperrins and around Glendun in Antrim, where the gold has originated in quartz veins in the Dalradian schists. As gold invariably occurs as native metal, it can be melted and cast into moulds, or because of its extreme malleability it can be worked into ornaments by hammering and beating.

EXCURSIONS

Previous page An aerial view of Scrabo Tower, the nineteenth-century monument on Scrabo Hill, near Newtownards, County Down. The hill consists of a Palaeogene dolerite sill capping sandstones of the Triassic Sherwood Sandstone Group.

Chapter 8

The west

The Ox Mountains and north Sligo

The area around the town of Sligo and bordering the shore of Sligo Bay and the southern part of Donegal Bay includes some of the most spectacular scenery found anywhere in Ireland and is also famous for its archaeological remains, particularly in the form of impressive megalithic monuments. Because of the association of the region with the poet W. B. Yeats, it is often referred to as Yeats country.*

In geological terms the area can be sharply divided into the metamorphic rocks of the northeast Ox Mountains, of problematical age, and the various (mainly sedimentary) rocks of Carboniferous age, ranging from 350 to 329 million years ago, which surround them (Fig. 8.1). The northeast Ox Mountains run in a narrow band with a roughly east-northeast trend from south of Sligo town via Lough Gill to around Manorhamilton in Country Leitrim. After the gneisses of Inishtrahull off the north coast of Donegal, and other gneisses in northwest Mayo, the northeast Ox Mountains are probably the next oldest rocks in Ireland. They are now mainly quartzo-feldspathic gneisses and were originally deposited as shallow marine sediments after 1500 and before 600 million years ago. At about 600 million years ago they were converted to gneisses deep in the crust. By about 475 million years ago this block was added to the Laurentian margin. The rocks of the Ox Mountains were further changed when the Iapetus Ocean narrowed and closed to form the Caledonian Mountains between 475 and 385 million years ago.

The metamorphic rocks of the northeast Ox Mountains

The northeast Ox Mountains form an elongate ridge of hills, generally 300–500 m high, south and east of Sligo and forming the southern shore of Lough Gill. The rugged skyline of the Ox Mountains is in marked contrast

---

* OS sheets 25 and 16, 1:50 000; Geological Survey sheet 7, 1:100 000; *Geology and landscape in Yeats country*, Daniel Tietzsch-Tyler, Geological Survey of Ireland (pamphlet).

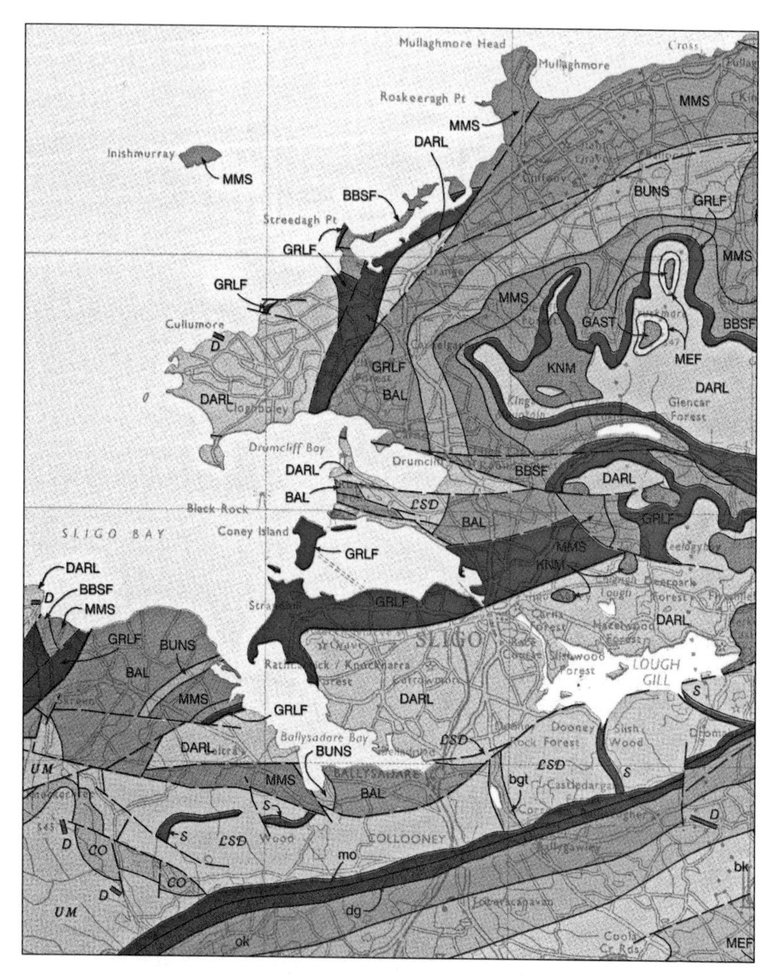

Figure 8.1 The northeast Ox Mountains and north Sligo (see p. 81 for the keys to this map; © Crown copyright).

to the flat-top hills north of Sligo, especially Ben Bulben, which are Carboniferous in age.

A convenient area to examine the gneisses of the Ox Mountains is on the south side of Lough Gill and in the Slishwood Gap G739315 to G748292. From the N4 road south of Sligo town, take the R287 to Ballintogher (G691329). Turn into the car-park and picnic area at Slishwood (G738315). From the car-park take the path to the left, which initially follows the shore of the lake, then turns into the woods, and then circles back to the car-park. The total walk is about 4 km. Beside the path at the lake shore are outcrops and blocks of grey and white banded gneisses (Fig. 8.2).

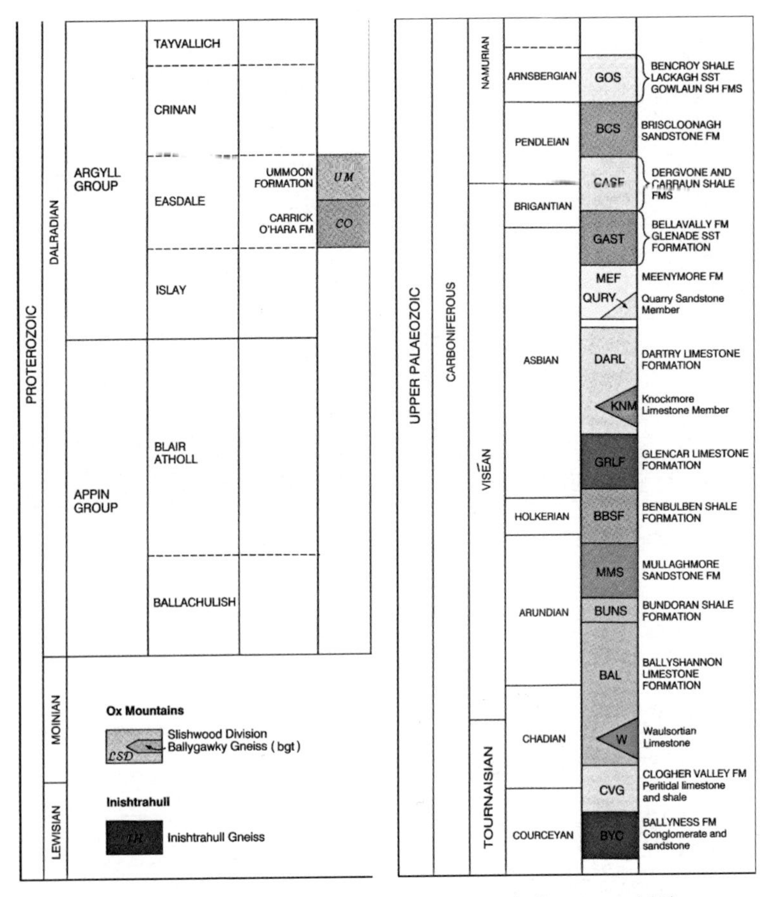

Figure 8.1 (continued) Keys for the map on p. 80. (© Crown copyright.)

These clearly show the contorted mineral banding that is characteristic of gneisses. The bands represent varying proportions of the minerals feldspar and quartz, with some mica. Quartz is grey and translucent, mica can be the pale form muscovite or the dark biotite, both showing a flaky structure, and the feldspar is pinkish in places.

Take the right-hand U-turn in the path at G750328. Several outcrops of gneiss occur along this section of the path, including a large boulder found on the south side of the path about 50 m after crossing a stream. This boulder shows excellent examples of tight folds in banded gneiss (Fig. 8.3). Follow the path back to the car-park. Follow the R287 road south from the car-park through the forest. This is the Slishwood Gap. The gap has been eroded along a broad band of the rock serpentinite, a mantle-derived rock, softer than the surrounding gneisses. This serpentinite was once mantle

Figure 8.2 Grey and white banded gneiss, south side of Lough Gill.

peridotite, thrust upwards from the mantle into the crust where the original hard minerals, pyroxene and olivine, were altered to a softer mineral, serpentine. The valley is noticeably more fertile than the surrounding hills, because of the magnesium-rich nature of the serpentinite minerals, compared to the acid soils derived from the surrounding gneisses. The serpentinite can be examined in a farm lane at G746294 on the west side of R287,

Figure 8.3 Tight folds in banded gneiss, Slishwood.

Figure 8.4 Serpentinite outcrop, Slishwood Gap.

just before the crossroads. The serpentinite is dark green and shows thin criss-crossing veins of the fibrous asbestos mineral chrysotile. Parking at this locality is limited and permission should be sought from the householder before visiting the outcrop (Fig. 8.4).

The Carboniferous rocks

Carrowmore Megalithic cemetery Proceed west to the N4 at Ballysadare; take the R292 west and follow the signs for Carrowmore Megalithic cemetery visitor centre (G662337). This area contains many megalithic monuments, and detailed information on their structure and significance is available at the visitor centre.

Knocknarea West of Carrowmore at the end of the Strandhill peninsula stands the hill of Knocknarea (G626346), 327 m high and capped by the highly visible stone cairn, traditionally supposed to be the tomb of first-century Queen Maeve of Connacht, but probably several thousand years older than this and linked to the burial monuments at Carrowmore. Knocknarea is accessible by a relatively easy walk that allows examination of the Carboniferous sediments that make up the hill, and the views from the summit over the whole Sligo Bay area are spectacular. Allow 40–45 minutes walking time from the car-park to the summit.

Park in the car-park on the southeast flank of the hill at G639339 and take the path signposted to the summit. The first 700 m are beds of blue-grey limestone, the Dartry Limestone, which forms the more spectacular slopes of Ben Bulben to the north. It formed as a lime mud in a tropical sea, about 350 million years ago in the Carboniferous. Within the limestone are nodules and bands of chert, a form of quartz or silica that is much harder than the calcite that makes up the limestone and so stands out as sharp edges and protrusions. Chert was used for the same toolmaking purposes as flint in the Cretaceous limestones of northeast Ireland; many of the graves in the area have yielded scrapers and other chert tools.

A turlough (periodic lake) can be seen in the field northeast of the bend in the track at G633344. The ephemeral nature of the lake is attributable to wide fluctuations in the water table between winter and summer. Exposures beside the path show bush-like coral fossils known as *Lithostrotion*, made up of branching tubes. Other fossils such as brachiopods, crinoids and bryozoans can be seen in the limestone outcrops along the path. Proceed to the summit area and the cairn of Maeve's Tomb, the largest monument in the area and thought to belong to the final phases of the Irish megalithic tradition (5000 years ago). It is enclosed by a bank 3 m wide and 0.2 m high, and surrounded by smaller structures, three being cairns. To the immediate north and south of the cairn are two large stones, interpreted as north and south markers. The cairn is clearly visible from the megaliths at Carrowmore, about 4 km away. The panoramic view from the summit shows the variation in landscape in the area. North and northeast of Sligo is the glacially derived U-shape valley containing Glencar Lough, and north of Glencar is the conspicuous hill of Benbulben, so closely associated with W. B. Yeats and his poetry. The sharp contrast between the grassy areas of limestone bedrock and the rugged heather-covered Ox Mountains is a fault line, and the contrast in landscapes is conspicuous (see Fig. 8.1).

Drumcliff church From Knocknarea follow the R292 through Strandhill towards Sligo. Further evidence of the earliest inhabitants of the area can be seen in the kitchen middens exposed on the bend at Culleenamore

(G611340). These consist of piles of oyster and mussel shells, up to 5 m thick, which indicate an abundance of food from the sea and the use of these sites over a very long period of time, probably from the Neolithic to the Early Iron Age. From Sligo take the N15 northwards to Drumcliff at G679420. Here in the Church of Ireland graveyard is the grave of W. B. Yeats, a poet with a strong sense of the Irish landscape. The headstone is limestone and inscribed with the epitaph:

Cast a cold eye
On Life on Death
Horseman pass by

The site also has an eleventh-century round tower and High Cross.

Benbulben and the Gleniff Horseshoe Road

For the best views of Benbulben proceed north on the N15 to the cross-roads at G656472, signposted to Ballinatrillick Bridge at G738503. This road passes the northern face of Benbulben (526 m high) and clearly shows the Dartry Limestone that forms the vertical cliffs of the upper part of the hill, whereas the lower slopes are composed of the Glencar Limestone, which is made up of thinly bedded pale limestones with interbedded darker shales (Fig. 8.5). Just west of Ballinatrillick Bridge take the road to the right (south), signposted to Gleniff Horseshoe. Gleniff is a natural amphitheatre, formed as a corrie by a valley glacier during the most recent ice age. In the backwall of the corrie there are several geological features. The left-hand or eastern side shows large-scale layering at an angle of

Figure 8.5 Benbulben with the Dartry Limestone overlying the Glencar Limestone.

about 45° to the horizontal. These are the layers of a mud mound formed on the bed of the Carboniferous sea and now consisting of fine-grain limestone. In places this limestone shows signs of karstic solution, which has created a system of cavities and caves. The large cave with a prominent arch high on the west side of the backwall and known as Diarmuid and Grainne's Bed is one such example (Fig. 8.6). The deep gash visible at the top centre of the headwall is a large solution hollow or swallow hole, which has been broken into and exposed as the headwall of the corrie eroded back. The mud-mound limestone overlies the Glencar Limestone (limestones and interbedded shales as seen at the base of Benbulben) and the junction is marked by a prominent springline where groundwater that has percolated through well jointed mud mound meets the first impermeable shale beds of the Glencar Limestone.

Some small openings on the backwall are small-scale mines for the mineral barytes. A larger deposit was worked until the late 1970s at Glencarbury some 2 km to the south. However, the road leading to the mine workings is currently closed to the public. Complete the circuit around the Horseshoe and proceed back to the N15 via Ballintrillick Bridge.

Streedagh Point (G637508) Proceed north along the N15 road to Grange and take the minor road to the west to Streedagh Point. Park at G637508. Walk northwest on the foreshore, towards the point. The exposed beds of the Glencar Limestone Formation contain a spectacular fauna comprising crinoid ossicles, brachiopods (*Productus*) and abundant corals (*Carinia* and *Lithostrotion*) (Fig. 8.7). A similar fauna can be examined at Serpent Rock at G567461 some 7 km southwest. Return to the N15.

Mullaghmore Head Proceed north on the N15 from Grange to Cliffany (G706536). Take the R279 northwards towards Mullaghmore Head. Park

Figure 8.6 Cave system known as Diarmuid and Grainne's Bed, Gleniff Horseshoe Road.

Figure 8.7 Coral and brachiopod fauna in the Glencar Limestone, Streedagh Point.

by the roadside at G698571 and walk down the steep path to the bay at the southern end of the headland and north across the wavecut platform. Here the Mullaghmore Sandstone Formation was deposited as a major river delta, built out into the tropical sea responsible for the limestones seen elsewhere in the area, from a landmass to the north. Interbedded shales and thick sandstones show internal structures such as cross bedding, graded bedding, and erosional channels with signs of burrowing and feeding trails (Fig. 8.8). These are particularly well developed on the undersides of beds in the cliff, which also show a cone-shape nautiloid, a creature related to the ammonites.

Extreme care must be exercised when walking on the wavecut platform, particularly during wet or windy weather.

South Donegal

The landscape of south Donegal shows marked contrasts between the ancient metamorphic rocks of the Slieve League Peninsula or the Lough Derg inlier, with their much more rugged scenery compared with the relatively gentle landscape around Donegal Bay.* This gentler scenery reflects the younger less-resistant sediments (shales, sandstones and limestones) of the Carboniferous age. In addition, the metamorphic upland areas

* OS sheets 10 and 11, 1:50 000; *Geology of south Donegal*, C. B. Long & B. J. McConnell, Geological Survey of Ireland.

Figure 8.8 Erosional channels in the Mullaghmore Sandstone, Mullaghmore Head.

mostly show features of glacial erosion, resulting in the sculpting of deep valleys along the lines of major structural faults (e.g. the Barnesmore Gap northeast of Donegal town). In contrast, the area around Donegal Bay shows signs of glacial deposition, with plenty of drumlins and boulder clay. The excursion reveals the contrasting lithologies of the older metamorphic rocks and the younger Carboniferous sediments (Fig. 8.9).

The main body of metamorphic rocks to be examined is the Dalradian rocks of the Slieve League to Glencolumbkille areas, west of Donegal town. These formed on a sea floor between splitting fragments of the supercontinent Rodinia about 700 million years ago. Originally sandstones, limestones and mudstones, the sediments were altered to schists, marbles and quartzites by mountain-building episodes during the Grampian orogeny about 475 million years ago. These Dalradian metasediments were then intruded 415–385 million years ago by the large granitic plutons that are such a major feature of the central and northern Donegal landscape.

Within the Dalradian sequence is a distinctive conglomerate deposit that was produced in a glacial marine environment, showing that, at that time, the northern part of Ireland was in high southern polar latitudes, in fact near the present Antarctic coastline.

Following the phase of granite batholith intrusion, the Devonian period was one of mainly rapid erosion under continental desert conditions and there are no Devonian sediments found in south Donegal. During the succeeding Carboniferous period, Ireland was inundated by a northward-spreading sea. Coarse-grain onshore deposits from the erosion

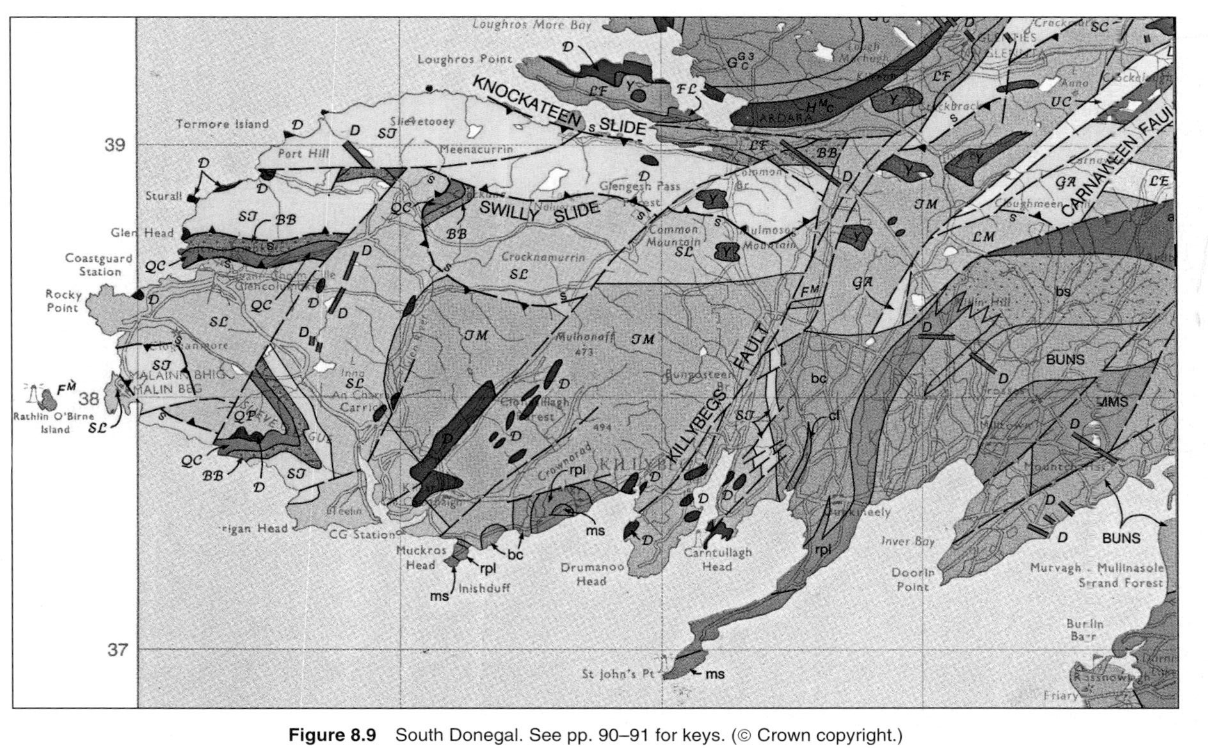

Figure 8.9 South Donegal. See pp. 90–91 for keys. (© Crown copyright.)

		GROUP	SUBGROUP	Ox Mountain County Sligo		Inishowen/Slieve League North County Donegal		South and Central County Donegal	
PROTEROZOIC	DALRADIAN	SOUTHERN HIGHLAND GROUP				GREENCASTLE GREEN BEDS	IB	SH	UPPER DALRADIAN (Undivided)
						INISHOWEN HEAD GRITS	IG		
						CLOGHAN GREEN BEDS	CI	CG	CROAGHGARROW FM ≡ Deele Group in Central Donegal
						FAHAN GRITS	FG	SG	SHANAGHY GREEN BED FORMATION
						FAHAN SLATES	FS	MY	MULLYFA FORMATION
		ARGYLL GROUP	TAYVALLICH			CULDAFF LIMESTONE	CL	AY	AGHYARAN FORMATION
			CRINAN			UPPER CRANA QUARTZITE	UC	KT	KILLETER QUARTZITE
						LOWER CRANA QUARTZITE	LC		
			EASDALE	UMMOON FORMATION	UM	TERMON PELITE	TM	TE / LM	TERMON PELITE LOUGH ESK PSAMMITE LOUGH MOURNE GRIT
				CARRICK O'HARA FM	CO	CRANFORD LST / SLIEVE LEAGUE FM	SC / SL	BG CO RF	BOULTYPATRICK GRIT CROAGHUBBRID PELITE REELAN FORMATION
			ISLAY			SLIEVE TOOEY QUARTZITE	ST	GA	GAUGIN QUARTZITE
						PORTASKAIG TILLITE	BB	BB	
		APPIN GROUP	BLAIR ATHOLL			GLENCOLUMBKILLE LST	QC	OC	OWENGARVE CALCAREOUS FM
						FINTOWN PELITE / GLENCOLUMBKILLE PELITE FORMATION	FI / QP		
						LOUGHROS GP AND Ur FALCARRAGH PELITE	LF		
						FALCARRAGH LIMESTONE	FL		
						Lr FALCARRAGH PELITE AND CLONMASS FM	FC		
			BALLACHULISH			ARDS QUARTZITE	AQ		
						ARDS BLACK SCHIST AND CREESLOUGH FM	AC		

Figure 8.9 continued Part of the key to the map on p. 89. (© Crown copyright.)

of Dalradian rocks to the north pass southwards into deeper-water limestones and shales, as can be found in the Sligo excursion. Mud and sand deltas were built in this warm shallow sea by rivers flowing from the north.

The crustal changes associated with the opening of the North Atlantic had a profound effect on the landscape of northeast Ireland and western Scotland, with the formation of vast lava plateaux such as in Antrim. However, even as far west as Donegal there are signs of the crustal tension that existed, with the intrusion of a swarm of dolerite dykes in many areas about 60 million years ago. A fine example of one of the thicker dykes can be seen at the Blind Rock on the north shore of Donegal Bay.

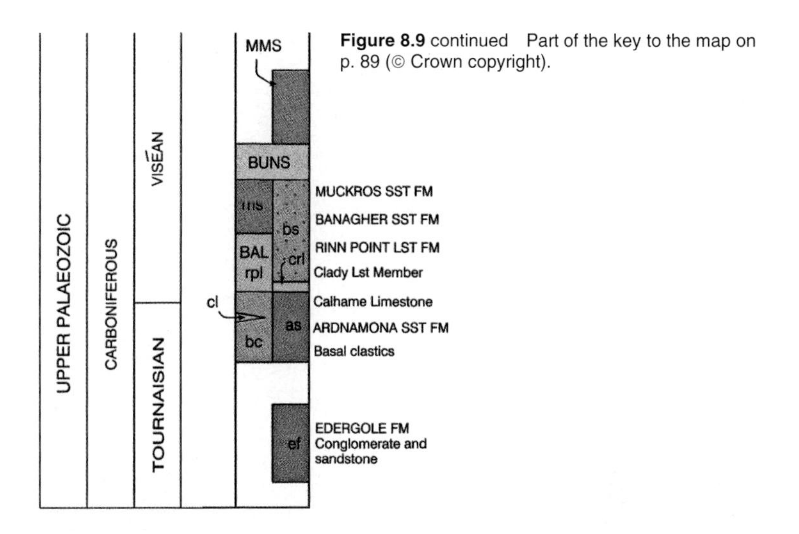

Figure 8.9 continued Part of the key to the map on p. 89 (© Crown copyright).

Donegal Bay–Slieve League–Glencolumbkille–Kiltyfanned Lough

Kiltyfanned Lough Take the R263 west from Donegal town to Carrick and take the road north to Meenancary. Proceed to G586884 between Kiltyfanned Lough and Lougheraherk, and park. A walk from here, around the low hill north of the road G582885, allows examination of the area in the sequence of its stratigraphical succession. The sequence in the area consists of the Glencolumbkille Limestone Formation, overlain by the Port Askaig Formation, overlain in turn by the Slieve Tooey Quartzite Formation (Fig. 8.9). The Glencolumbkille Limestone consists of a lower part, the Glen Head Schist Member, which occurs as interbedded quartzites and pelites, vertically bedded (Fig. 8.10), passing up into the Glencolumbkille Dolomite Member, which shows conspicuous marble beds, weathering brown or yellow, interbedded with thin sandy or quartzite beds (Fig. 8.11).The sequence is in fact overturned structurally, with the oldest members the Glenhead Schist and Glencolumbkille Dolomite exposed at the top of the hill, the younger rocks occurring farther down the hill. Stratigraphically above the Dolomite Member is the Port Askaig Formation. There is a gradual transition from the Dolomite Member to the boulder bed or tillite of the Port Askaig Formation. Tillite is the term for a metamorphosed till or boulder clay. The rock is silty, with occasional marble clasts becoming semi-pelitic with granite clasts. The succession represents an upward-shallowing sequence and is interpreted as a consequence of ice-sheet melting and subsequent continental rising. Siliceous interbedded units are probably meltwater or tidal deposits. Granite clasts within the fine-grain matrix are up to 50 cm in diameter in places (Fig. 8.12).

Figure 8.10 Glenhead Schist member, interbedded quartzites and pelites, Kiltyfanned Lough.

Proceed down slope to the road along the shore of Kiltyfanned Lough and go east. The Slieve Tooey Quartzite Formation occurs in exposures along the north side of the road.

Continue to outcrops south of the road at G589884, 200 m west from Lougheraherk school (Fig. 8.13). Here, on a prominent south-facing slab approximately 200 m south of the road, a deformed polygonal pattern has

Figure 8.11 Glencolumbkille Dolomite, Kiltyfanned Lough.

Figure 8.12 Granite clast in the Port Askaig Formation, Kiltyfanned Lough.

Figure 8.13 Ice-wedge casts, Port Askaig Formation, Lougheraherk school.

been interpreted as ice-wedge casts, a phenomenon observed in contemporary tundra areas. The cut peat bogs here show the preserved trunks and root systems of the forests that covered the area before the generation of the peat deposits.

The Port Askaig Formation can also be examined during the north Donegal excursion (pp. 103–104).

From Kiltyfanned proceed to Glencolumbkille and take the road out of the village to the northwest to Skelpoonagh Bay at G521861. Here on the wavecut platform are clearly exposed minor folds of thin quartzite beds in psammites of the Glen Head Schist Member (Fig. 8.14).

The area around Glencolmkille is rich in archaeological remains from the Neolithic, with many fine examples of court cairns, standing stones and portal dolmens. There is also an important association with Saint Columba and the early Christian Church in Ireland; many of the standing stones were adopted to Christian use and form the stations of a pilgrimage called the Turas Colmkille, which is performed each year on 9 June, Saint Columba's day. The standing stones and dolmens are clearly marked and there is a folk museum.

Slieve League Proceed to Carrick (G593790) via the R263 from Glencolumbkille and take the road south from Carrick along the west side of Teelin Bay to the viewpoint at Carrigan Head (G559757). Here Slieve League forms the highest cliffs in Ireland, with the summit of Slieve League itself at 595 m. The cliffs are mainly quartzites of the Slieve Tooey Quartzite Formation (Fig. 8.15). Care should be taken, particularly in poor weather.

Figure 8.14 Folding in the Glen Head schist, Skelpoonagh Bay, Glencolumbkille.

Muckros Head and Shalwy Retrace the route to Carrick and proceed east on R263 to Kilcar. Take the road south to Muckros Head at G618736. Here the Carboniferous Muckros Sandstone Formation forms the cliffs and wavecut platform, comprising flat-lying calcareous sandstones (Fig. 8.16). To the east of Muckros Head, near Shalwy (G646748) the Carboniferous basal conglomerate of the Rinn Point Limestone Formation is seen to overlie unconformably the schists of the Dalradian Termon Formation (Fig. 8.17). Access to this locality is possible only at low tide, and parking on the nearby road is very restricted.

Blind Rock Dyke An excellent example of the Palaeogene dyke swarm in Donegal can be examined on the foreshore at Inver (G851745) on the north shore of Donegal Bay, 4 km southwest of Mountcharles. The dyke is subvertical and about 30 m wide. It is composite in form: it contains several varieties of dolerite including an amygdaloidal form. It trends northwest–southeast and is intermittently exposed over a distance of around 3 km. Its thickness and composite nature suggest it may have been a feeder to lava flows on the surface, but like many such dykes to the west of the main outcrop of the Antrim lavas there is no direct evidence of such a role. Proceed to Donegal town.

Figure 8.15 Sea cliffs at Slieve League, mainly quartzites of the Slieve Tooey Quartzite Formation.

Figure 8.16 Cliffs of Carboniferous Muckros Sandstone at Muckros Head.

Figure 8.17 The basal conglomerate of the Carboniferous Rinn Point Limestone Formation, resting unconformably on the schists of the Dalradian Termon Formation, near Shalwy, south Donegal.

North Donegal

The geology of north Donegal is overwhelmingly dominated by Dalradian metasediments and their associated granitic intrusions. The area is also clearly marked by the processes of glacial erosion and deposition; these processes mostly account for the long sea inlets such as Lough Swilly and Mulroy Bay, which result in a very complicated coastline. In plate-tectonic terms the structural geology of north Donegal is dominated by the effects of the closure of the Iapetus Ocean in Silurian times, with the compressive forces involved producing complex folding and faulting with large-scale thrust faulting moving large blocks of rock considerable distances. These tectonic forces were also melting country rock at depth to produce granitic magmas to form the series of plutons that make up the suite of Donegal granites, which are such a major factor in the central and northern Donegal landscape. This phase of intrusion lasted until Devonian times; the Devonian is also represented in the unconformable occurrence of conglomerates of this age lying on metamorphic rocks on the east side of the Fanad Peninsula, near Portsalon.

This excursion aims to illustrate the regional tectonics and the lithologies of some of the granite plutons and the associated country rocks. See Figure 8.18 for the relevant part of the 1:250 000 geology map.

Horn Head and the Dunfanaghy area
For this section of the excursion, allow 3–4 hours.

Horn Head slide The Horn Head slide is at Mickey's Hole, opposite Harvey's Rocks (B988400), about 4 km northwest of the village of Dunfanaghy (C017373). The locality is accessible at all states of the tide. The horizontal compressive forces that closed the Iapetus Ocean and built the Caledonian Mountains folded and faulted the sandstone and siltstone beds, and intensely deformed them. In some cases the horizontal forces moved large blocks along low-angle fault planes called slides, so that large volumes of rock were moved over adjacent rocks. At Mickey's Hole on Horn Head the precise succession is still under discussion, but that proposed by the recent mapping of the Geological Survey of Ireland (Long & McConnell 1997) is the Ards Pelite Formation, overlain by the Ards Quartzite Formation:

Ards Quartzite Formation
Ards Pelite Formation.

The quartzite is whitish, the pelite is dark grey or black. At this locality the dark Ards Pelite appears to have been thrust over the younger pale Ards Quartzite (Figs 8.19, 8.20). The quartzite is conspicuously platy close to the faulted contact. To visit this locality, park in the small disused quarry at B997402 and take the track to the west, which starts 100 m along the road to the north, on the left. The track to the coast passes a megalithic tomb at B994403.

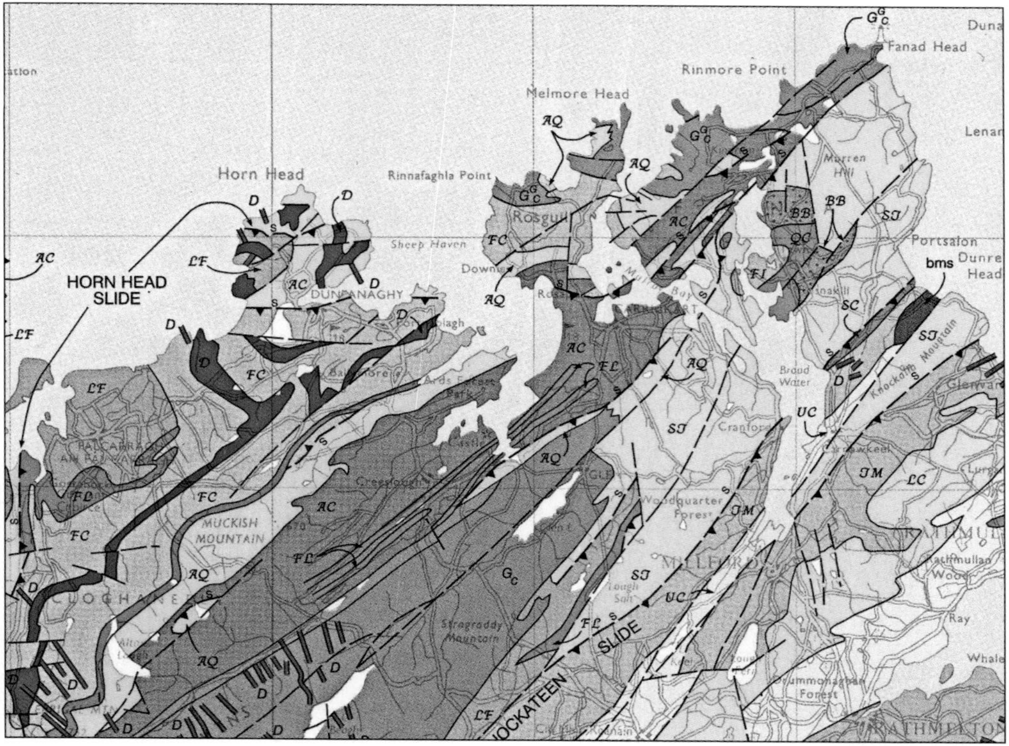

Figure 8.18 North Donegal (GGC Fanad granite; GC Main Donegal Granite. See p. 100 for the keys to this map. (© Crown copyright.)

		GROUP		
PROTEROZOIC	DALRADIAN	SOUTHERN HIGHLAND GROUP	GREENCASTLE GREEN BEDS	IB
			INISHOWEN HEAD GRITS	IG
			CLOGHAN GREEN BEDS	CI
			FAHAN GRITS	FG
			FAHAN SLATES	FS
		ARGYLL GROUP	CULDAFF LIMESTONE	CL
			UPPER CRANA QUARTZITE	UC
			LOWER CRANA QUARTZITE	LC
			TERMON PELITE	TM
			CRANFORD LST	SC
			SLIEVE LEAGUE FM	SL
			SLIEVE TOOEY QUARTZITE	ST
			PORTASKAIG TILLITE	B
		APPIN GROUP	GLENCOLUMBKILLE LST	QC
			FINTOWN PELITE	FI
			GLENCOLUMBKILLE PELITE FORMATION	QP
			LOUGHROS GP AND Ur FALCARRAGH PELITE	LF
			FALCARRAGH LIMESTONE	FL
			Lr FALCARRAGH PELITE AND CLONMASS FM	FC
			ARDS QUARTZITE	AQ
			ARDS BLACK SCHIST AND CREESLOUGH FM	AC

Figure 8.18 continued Keys to the map on p. 99. (© Crown copyright.)

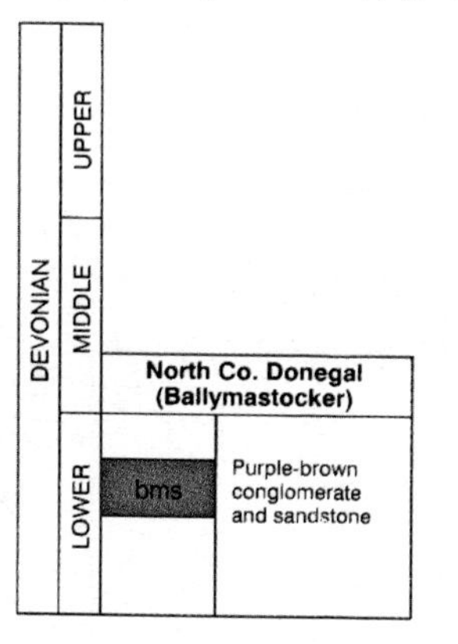

Figure 8.19 The Horn Head slide at Mickey's Hole, showing the dark Ards Pelite thrust over the younger Ards Quartzite to the left of the field of view.

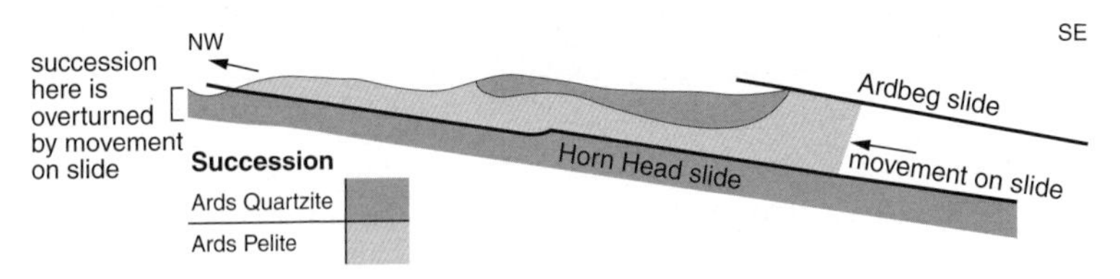

Figure 8.20 Cross section through the Horn Head slide at Mickey's Hole (after Long & McConnell 1997).

Breaghy Head Return to Dunfanaghy and proceed east on road N56. Take the road to the left (northeast) at C047364, the first on the left after the post office at Rockhill, and go to the bend on the road (C057376); there is limited parking on the verge here. Proceed westwards to the coast (C053377). Here the Breaghy Head slide can be seen bringing rocks of the Sessiagh–Clonmass Formation into contact with metadolerite. The former are mainly quartzites, with some calcareous weathering patches (Fig. 8.21). The cliff face to the south shows spectacular recumbent folding, where the limbs of the folds are almost horizontal and parallel to each other (Fig. 8.22). The outcrop is best seen at low tide.

Lackagh Bridge From Breaghy Head return to the N56 and drive to Creeslough (C058307) and take the R245 to near Lackagh Bridge to examine excellent examples of mullions in quartzite at C099317 (Fig. 8.23). The term "mullion" is used to describe the striking lineation seen occasionally in

Figure 8.21 Quartzites of the Sessiagh–Clonmass Formation in contact with metadolerite along the Breaghy Head slide.

folded quartzites and considered similar in appearance to the vertical bars separating the sections of a window.

Doe Castle, en route to Lackagh Bridge, is a fine example of an early sixteenth-century castle and is worth a visit if time permits.

Barnes Gap From Lackagh Bridge return to the N56 and head south from Creeslough. Approximately 1.5 km south of the road bridge over the River Owencarrow at C084259 is a large layby on the west side of the road. Along the east side of the road at this point is a disused railway line. Cross the road from the layby to the railway line and walk north. In cuttings along

Figure 8.22 Recumbent folding at Breaghy Head.

Figure 8.23 Mullions in the Ards Quartzite near Lackagh Bridge.

this line there are good exposures of the Main Donegal Granite available for examination and sampling. At C083264, just south of where the old railway line crossed the road, are enclaves of pelite, metadolerite and marble within the granite.

Fanad Peninsula

For this section of the excursion, allow 3–4 hours. From Barnes Gap, proceed to Milford (C192409) at the southern end of Mulroy Bay and then north along the R246 on the east side of Mulroy Bay, through Carrowkeel (Kerrykeel) to Croaghan Hill at C192409, via Tawney (also called Tamney). Limited offroad parking is available here 50 m along the road to Portsalon.

The Port Askaig Formation The Port Askaig Formation, a glacially derived conglomerate or tillite occurs extensively in the area around Croaghan Hill. This rock can also be examined in the Kiltyfanned Lough area of south Donegal (pp. 91–94) and can be correlated with similar occurrences in Scotland (Port Askaig is on the island of Islay off the west coast of Scotland). It occurs stratigraphically within the Dalradian succession and it formed at a time when Ireland was part of supercontinent Rodinia 600 million years ago, and located in polar latitudes in the southern hemisphere.

The formation consists of a sequence of **diamictites**, the name given to the deposits dumped by a glacier at its base or margin or end. These would normally form moraines consisting of unsorted sediments varying from clay-size particles to large cobbles and boulders. The blocks or clasts can

Figure 8.24 Granite clasts in the Port Askaig tillite at Croaghan Hill, Fanad Peninsula.

also vary in shape from angular to rounded. Diamictites of the Port Askaig Formation show evidence that deposition was at sea level. Subaerial and shallow submarine deposits are present, as well as ice-rafted debris from shelf ice, dropped off shore from melting icebergs. The matrix of the diamictites on Fanad ranges from a pale buff dolomitic pelite schist to a darker grey psammitic schist. The clasts at Croaghan are common and range in size from 1 cm diameter to boulders of more than 20 cm in diameter. They are frequently granitic in composition, with dolomite, quartzite and vein quartz also occurring (Fig. 8.24). Interbeds of sandstones and siltstones occur between the diamictites and can be used to indicate the orientation of the bedding of the rock.

About 100 m along the road from the junction, towards Kindrum, is a gate to the outcrops above the road. These crags and those towards the summit of Croaghan Hill provide good examples of the tillite. The outcrops at the roadside are metadolerite. A traverse in a generally northeasterly direction from the road junction will pass through the middle section of the formation. Bedding appears to be dipping 20–30° northeast.

Lough Nagreany Proceed north to Kindrum and take the road to the west around Kindrum Bay to Lough Nagreany at C148416. The Fanad Granite occurs mainly on the northern end of the Fanad Peninsula, but also on the northern end of the Rosguill Peninsula to the west. Rocks at Melmore Head on Rosguill are an intricate mix of country rock and granite magma (migmatite), and form a transition zone between the country rock and the main granite mass. On Fanad this migmatite occurs around Lough Nagreany.

There is limited roadside parking here. Access to the lough is via a gate. Ards Quartzite crops out by the roadside. Follow the track parallel to the lough shore for 200 m to low crags. These are part of the migmatite zone of the Fanad Granite and they show clear examples of a dark pelitic country rock, as well as quartzite and vein quartz fragments (Fig. 8.25). This is referred to as the xenolithic variety of the granite on the geological map. Xenoliths are inclusions of other rocks found within igneous rocks.

Doaghmore Strand Follow the track in a general northerly direction across a stream and through sandhills to the southwest end of Doaghmore Strand at C138419. Here is the contact zone between the Fanad pluton and the country rock, which here comprises the Ards Quartzite and schists of the Sessiagh–Clonmass Formation. The effect of the intrusion of the granite on the country rock has been to thermally alter the country rock, with the formation of new minerals in response to the higher temperatures. Near the contact, small black crystals are found in the schists. These are probably andalusite, an aluminium silicate mineral, the presence of which indicates a relatively high-temperature contact metamorphism of the schists (Fig. 8.26). The outcrop is approximately 150–200 m northwest of the concrete end-cover of a land drain at the high-water mark on the southwest end of Doaghmore Strand. Excellent examples of the xenolithic facies of the Fanad Granite can be seen at the northwest end of the strand, and the Main Fanad Granite can be examined on the coast on the west side of Ballyhiernan Bay to the northeast of Lough Nagreany. Park at the car-park at C174447 and examine the coastal outcrops 100 m to the northwest.

Portsalon and the Old Red Sandstone

From Ballyhiernan Bay go east and follow the signs to Portsalon from the post office at Magheradrumman (C208450). Leave Portsalon on the R246 and proceed to the car-park at the southeastern end of Ballymastocker Bay at C250378. Cross the beach to outcrops near the high-water mark.

One of the unexpected findings of the Geological Survey of Ireland's mapping of the Fanad Peninsula in 1885 was of Devonian (Old Red Sandstone) rocks at Ballymastocker Bay, near Portsalon. This occurs as an outlier, a body of younger rock entirely surrounded by older rocks, in this case Dalradian. The connecting portion of Devonian has been removed by erosion, the nearest Devonian rocks are in Counties Fermanagh and Tyrone, some 75 km away. These Devonian rocks on Fanad have been preserved in a downfaulted block and in places inland the contact with the underlying Dalradian rocks is a good example of an unconformity as discussed earlier (see Fig. 6.5). The time difference between deposition of the local

Figure 8.25 The migmatite zone of the Fanad granite at Lough Nagreany, Fanad Peninsula.

Figure 8.26 Andalusite locality in schists of the Sessiagh–Clonmass Formation at Glinsk on the Fanad Peninsula.

Dalradian quartzites and the Devonian is about 300 million years. The rocks are conglomerates with coarse sandstone lenses, containing boulders of quartzite, schists, basic and acid igneous rocks, often with a high degree of rounding and up to 1 m in diameter (Fig. 8.27). The next outcrops along the beach to the southeast are of the Slieve Tooey quartzite that forms the nearby Knockalla Mountain, although here the contact is a fault.

Figure 8.27 Devonian conglomerate, Ballymastocker Bay, Portsalon, Fanad Peninsula.

Chapter 9

Fermanagh and Tyrone

County Fermanagh

The geology of the area covered in this excursion, to the south and west of Lower Lough Erne and south as far as Cuilcagh Mountain, is dominated by rocks of Carboniferous age. Many of these rocks are limestones, and the fact that such rocks dissolve in rain water results in some unique features in the geology of Fermanagh, such as the underground drainage and caves at Marble Arch and other features of a karst landscape (pp. 45–47). The Carboniferous is characterized by the gradual flooding of the earlier desert sediments of the Devonian period, as a warm shallow sea moved northwards, containing a rich variety of life forms, including shellfish and corals. In this excursion many of the features of karst landscape are examined, and also the variety of limestones that occur in the area (Fig. 9.1).*

Lower Lough Erne

Take the A46 road northwest out of Enniskillen (H240440) towards Beleek. Approximately 10 km along this road at Carrickreagh Bay (H174521) is a disused quarry on the west side of the road (Fig. 9.2). These are well bedded limestones of the Ballyshannon Formation (Fig. 9.1). At Blaney Quarry approximately 1 km northwest along the road at H167525, similar coarse-grain pale-grey limestones occur. Within these limestones occasional beds contain brachiopods, solitary corals and colonies of *Syringapora* coral.

The Ballyshannon Limestone Formation viewed here is one of the oldest of the Carboniferous units seen on the excursion. If time permits, follow the A46 northwest along the south shore of Lower Lough Erne to the spectacular Cliffs of Magho at H040560. The contact between the older Glencar Limestone and the overlying Dartry Limestone Formation is seen in the cliff exposures and is marked by a change in slope angle and vegetation

* OS 1:50 000 sheet 17, Lower Lough Erne, sheet 26, Lough Allen; GSNI Derrygonnelly and Marble Arch, sheets 43, 44, 56; additional geological information, *Geology of the country around Derrygonnelly and Marble Arch*, I. C. Legg, T. P. Johnston, W. I. Mitchell, R. A. Smith (GSNI 1998).

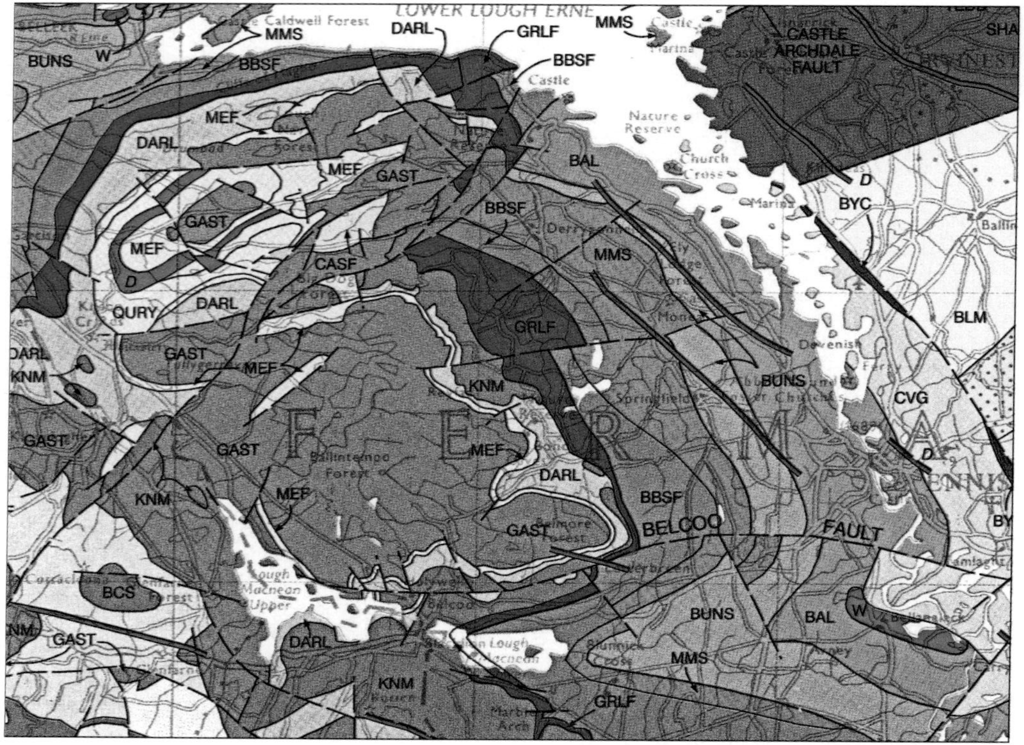

Figure 9.1 The Lower Lough Erne area. See p. 111 for keys to this map. (© Crown copyright.)

FINTONA BLOCK			
County Fermanagh		County Tyrone	
North of Killadeas-Seskinore Fault		South of Tempo-Sixmilecross Fault	
SHAN / A / TEDD	SHANMULLAGH FM Red sandstone, siltstone and mudstone Andesite (A) TEDD FM Conglomerate and sandstone	RSF	RAVEAGH SST FM
		GOOF	GORTFINBAR CONG FM
	BHAM		Barrack Hill Andesite Member
		SMSF	SHANMAGHERY SST FM

Curlew Mountains County Sligo		North Co. Donegal (Ballymastocker)	
kw	KEADEW FORMATION Sandstone, conglomerate and lavas	bms	Purple-brown conglomerate and sandstone

Figure 9.1 continued Keys to the map on p. 110 (© Crown copyright).

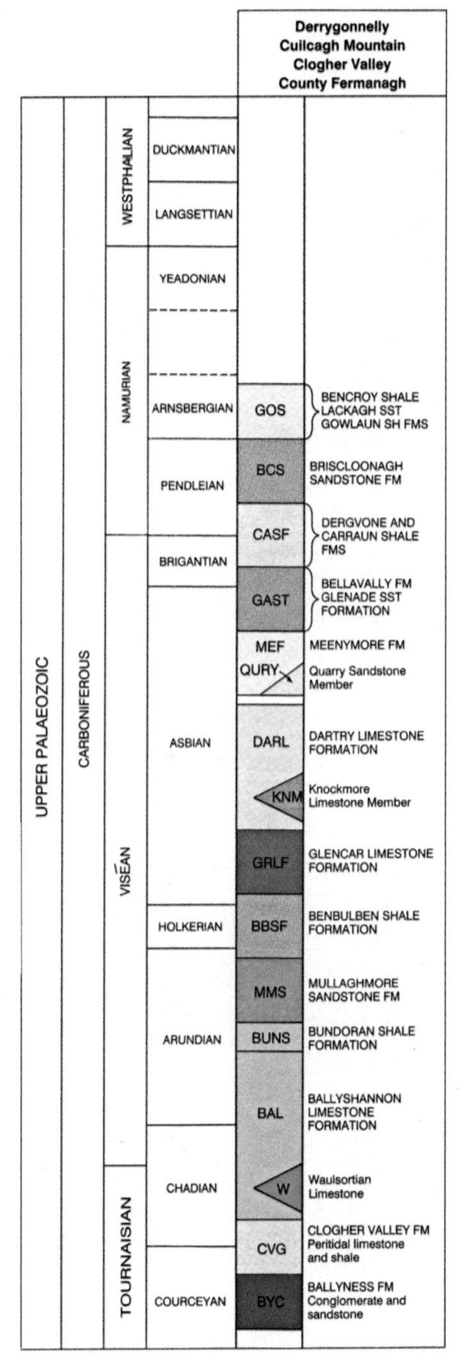

Figure 9.2 Limestones of the Ballyshannon Formation, Carrickreagh quarry, Lower Lough Erne.

and the occurrence of a spring line. Alternatively, proceed directly to Slisgarrow Quarry at H015517 on the north side of the road between Derrygonnelly and Garrison. Parking is available in a layby 100 m to the west of the quarry on the south side of the road. In the quarry the exposed upper contact of the Palaeogene (Tertiary) Garrison sill and the overlying thin-bedded shale and limestone of the Meenymore Formation can be examined (Fig. 9.3). The Garrison sill is olivine dolerite intruded into the western end of the Glen Syncline, which trends northeast–southwest

Figure 9.3 Upper contact of the Garrison sill against shale and limestone of the Meenymore Formation, Slisgarrow quarry.

(Fig. 9.1). Both the upper and lower contacts with the associated sediments can be seen in this quarry, with the 3–5 m thick dolerite sill showing crudely developed columnar jointing in places. The sediments above and below the sill have been thermally altered to hornfels in places.

Cladagh River, Cladagh Glen and Marble Arch caves

Proceed from Garrison to Belcoo along the B52 and thence to Blacklion (in the Irish Republic), and from Blacklion take the road to the southeast along the south shore of Lough Macnean Lower in the direction of Florencecourt. This road crosses the Cladagh River at Cladagh Bridge (H128358), where parking is available at the bridge.

Cladagh Glen leads to the show caves at Marble Arch and provides a readily accessible examination of the contact between the Glencar Limestone Formation and the overlying Dartry Limestone. There are also opportunities to examine resurgent drainage where three underground rivers emerge from the Marble Arch cave system as a single river, the Cladagh. From the car-park to the visitors centre at the Marble Arch caves is approximately 1.6 km. Follow the path up the glen on the east side of the river. At Cascades (H123350) is an impressive stepped waterfall resurgence of the Prod's Pot Cascades rising cave system (Fig. 9.4). Up stream from this point on the west bank of the river is the contact between the Glencar Limestone and the overlying Knockmore Limestone Member of the Dartry Limestone Formation (Fig. 9.5). The river here can be dangerous if water levels are high after heavy rain, and extreme caution is required if crossing the river to examine the contact. About 20–30 m up stream from

Figure 9.4 Resurgent drainage system at Cascades, Cladagh Glen.

Figure 9.5 Contact between the Glencar Limestone and the overlying Knockmore Limestone, Cladagh River, Cladagh Glen. © Crown copyright; reproduced by permission of the Director, Geological Survey of Northern Ireland.

this contact is the Marble Arch and it is essential that visitors keep to the paths and walkways provided. The Marble Arch rising marks the resurgence of the three rivers that flow off Cuilcagh Mountain to the south into the Marble Arch cave system. Here they emerge as the River Cladagh. The Marble Arch is a natural bridge spanning the Cladagh, being the remnants of a cave roof after collapse. The round hole to the left of the arch is called a **phreatic tube** and was formed by a small underground stream dissolving the limestone (Fig. 9.6). The Dartry limestone is relatively pure calcium carbonate and readily forms caves by solution. The underlying Glencar Limestone is impure and cave formation is limited, so the rising here occurs where the river meets the Glencar Limestone and cannot easily cut down into it.

Follow the steps up through the nature reserve to the visitors' centre at the Marble Arch caves, passing several information boards erected by the Environment and Heritage Service. An interesting and informative display is to be found in the visitor centre, along with refreshments, toilets and access to the show caves. The centre closes from October to March and access should be checked on their website. Retrace the route to the car-park at Cladagh Glen. Allow 1–2 hours for the round trip without a visit to the show caves. Allow up to an extra 2 hours depending on availability and guided-tour timings if a cave trip is to be included.

Figure 9.6 Marble Arch over the River Cladagh. © Crown copyright; reproduced by permission of the Director, Geological Survey of Northern Ireland.

Gortalughany Quarry and the Legacurragh Gap

From Cladagh Bridge proceed east towards Florencecourt House and the A32 Enniskillen–Swanlinbar road at H193356. Turn south along this road for approximately 5 km to H195302 and take the signposted road (Brown Signpost) to Gortalughany scenic drive. Park on roadside at H168303 near a large disused quarry (Fig. 9.7). Here there are good exposures of thin and thick-bedded Dartry Limestone, with thin calcareous shales and conspicuous layers of blue-black chert. On the downslope side of the road near the

Figure 9.7 Dartry Limestone near Gortalughany.

Figure 9.8 The Legacurragh Gap.

quarry is a disused lime kiln. Proceed up the road to the car-park and viewpoint at the top of the Gortalughany Road at H168300 and park. Proceed west along the road to a junction with a track closed by a gate at H165300. Proceed north along the track, which is used by turf or peat cutters. The Legacurragh area around H158308 shows karstic features in the Dartry Limestone. The Legacurragh Gap (Fig. 9.8) is a steep dry valley running northeast below a cliff. It was probably formed by surface drainage during permafrost conditions at the end of the most recent ice age. Glacial meltwater flowing out in torrents from the edge of the ice carved the valley in the frozen ground, but following the melting of the ice and the end of permafrost conditions the drainage would have reverted to being subterranean.

Around Legacurragh are sinkholes or pots, often marked by vegetation more luxuriant than the surrounding heather. Adjacent to the path at Legacurragh there are areas of **limestone pavement** developed where glaciation has stripped off the superficial deposits and exposed the limestone bedrock. This has then been eroded along vertical joints to produce water-widened cracks or fissures, known as **grykes**, with upstanding blocks of limestone between them, known as **clints** (Fig. 9.9). There are frequent exposures of colonial corals on the limestone pavement around this locality.

Retrace your route back to the car-park (20–30 minutes) and return to Enniskillen. Allow 1–2 hours for the round trip for this part of the excursion.

Figure 9.9 Limestone pavement with clints and grykes, Legacurragh.

County Tyrone

County Tyrone exhibits a wide range of geological environments, from the Dalradian metamorphic rocks of the Sperrin Mountains in the north through the Caledonian igneous rocks of the Tyrone **ophiolite** complex in the south, with rocks of possible Moinian age in the Central Tyrone inlier. Ophiolites are a sequence of deep-sea sediments, pillow lavas, dykes and gabbros that formed on the ocean floor at constructive plate margins. Those ophiolites found on land, such as in Tyrone, are slices of ancient ocean floor that were caught up in plate collisions and thrust up, or **obducted**, onto continental crust. In addition, the area is crossed by the Omagh Fault, one of the most important structural discontinuities in the north of Ireland, thought be one of the terrane boundaries in the intermediate zone between those with Gondwana affinities to the south and those with Laurentian affinities to the north (see pp. 40–41 and Fig. 5.9).

The excursion aims to examine some of the component parts of the Tyrone ophiolite in association with island-arc volcanic rocks that were erupted along the northern margin of the Iapetus as the ocean was progressively closing. The Omagh Fault will also be examined, as well as key lithologies in the Dalradian succession to the north. Allow 5–6 hours for the full excursion.

The Dalradian succession

Butterlope Glen (H493948) From Omagh (H450725) take the B48 northwards to Gortin and Plumbridge. Approximately 5 km north of Plumbridge at H490945 is the road through Butterlope Glen, running southeast from the B48 (Fig. 9.10). The glen is a deeply incised glacial meltwater channel that cuts across the regional strike and structure of the area. At the end of the most recent ice age, about 13 000 years ago, meltwater from the retreating Sperrins ice sheets was prevented from draining away and filled the Glenelly River valley to the south of Butterlope. As water levels rose, they eventually found an escape route by incising a narrow deep channel along what is now Butterlope Glen. The rocks at the northern (bottom) end of the glen are part of the Dungiven Formation, the top of the Tayvallich Subgroup of the Argyll Group of the Dalradian. The southern (top) end of the glen is in the succeeding Dart Formation of the Southern Highland Group (Fig. 9.11). This means that, although the geological succession dips to the north, it is structurally inverted, that is the beds have been turned upside down. Therefore, progressively older beds occur down dip towards the north of the glen, whereas in a normal succession the beds become younger down dip.

Take the Butterlope Road from the north end of the Glen to the fork at the south end, then take the east fork (the Bradkeel Road) and park 100 m along in a rough layby on the south side of the road. Work back down the glen; the best exposures are seen on the east side.

At the hairpin bends at the north end of the glen are the youngest rocks in the area, silvery-grey amphibolite schist of the Glenga Member of the

Figure 9.10 Butterlope Glen, County Tyrone.

Figure 9.11 County Tyrone. See pp. 120–121 for keys to this map.

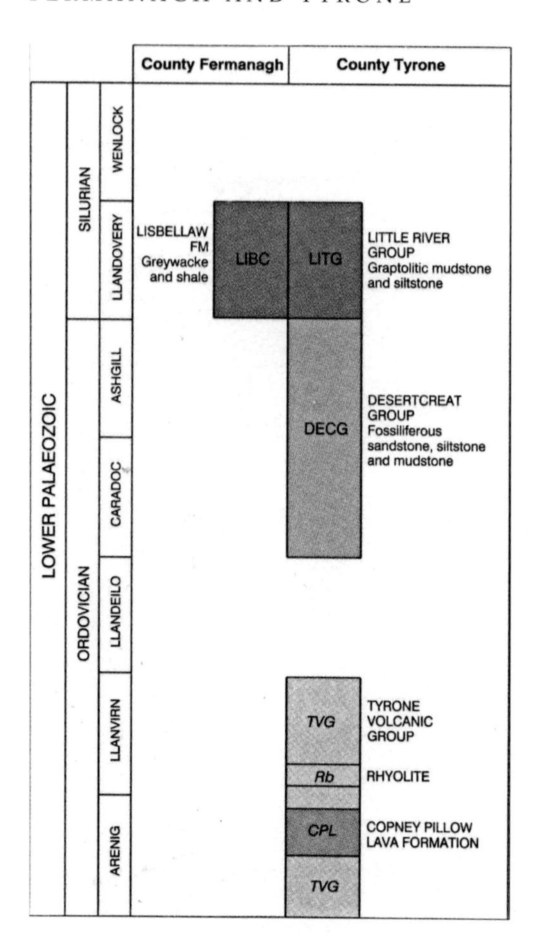

Figure 9.11 continued
First part of the key to the map on p. 119. (© Crown copyright.)

Dart Formation. These give way northwards to mixed interbedded quartz–biotite psammites and green amphibolite schist. Typical limestones of the Dungiven Formation are exposed in disused quarry pits near the roadside, about 200 m from the northern entrance to the glen (Fig. 9.12). At the north end of the glen on the crags behind the farmhouse at H491954 are several outcrops of highly deformed volcanogenic rocks, including stretched pillow lavas and interbedded amphibolite schist (Fig. 9.13).

High on the east slope, half way down the glen, near a concrete cistern on the east of the road, are thick units of pale buff quartzite forming a large flat-lying downward-facing fold (Fig. 9.14).

Craig Pillow Lavas (H523981) From the layby at Butterlope Glen proceed north again to the B48 and travel north approximately 1.5 km and take the

	GROUP	SUBGROUP	(North) Co Londonderry	Sperrin Mountains		(South) County Tyrone
DALRADIAN	SOUTHERN HIGHLAND GROUP				MG	MULLAGHCARN FORMATION
			LONDONDERRY FORMATION	LY	GW	GLENGAWNA FORMATION
			BALLYKELLY FORMATION	BF	GY	GLENELLY FORMATION
			CLAUDY FORMATION	CF	DT	DART FORMATION
	ARGYLL GROUP	TAYVALLICH		DN		DUNGIVEN FORMATION
		CRINAN		NF		NEWTOWNSTEWART FORMATION
		EASDALE				
		ISLAY				

LITHOLOGICAL EXPLANATION

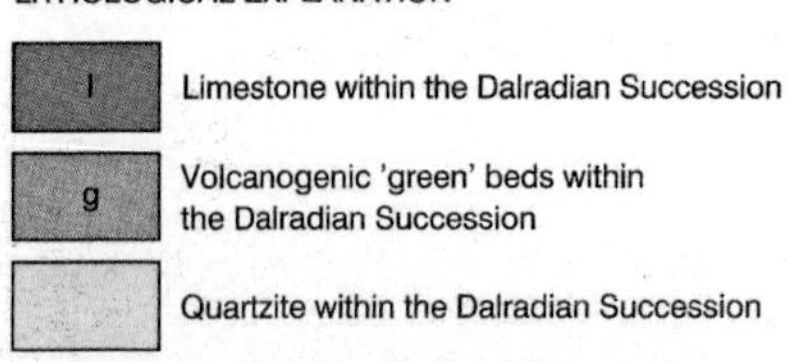

NB Formation names within the Dalradian frequently contain a reference to the lithology. Where no lithological term is used, the formations are mixed and consist of several different metamorphic rock types, none of which is dominant

Figure 9.11 continued Second part of the key to the map on p. 119. (© Crown copyright.)

first road on the right to Craig at H523981. There is room for limited parking of cars near the small group of houses and outbuildings. At the time of writing, one of these was occupied and permission should be sought here before examining the crags to the north of the road. Here are excellent examples of pillowed basalt lavas of the Dungiven Formation, the same age as the rocks examined previously in Butterlope Glen. Despite regional

Figure 9.12 Limestones of the Dalradian Dungiven Formation, Butterlope Glen.

metamorphism the pillows have retained many of the features characteristic of younger pillow-lava formations (Fig. 9.15). Individual pillows are up to 1 m in diameter and in places the original chilled margin of the pillow can be identified, along with gas bubbles or vesicles and radial-fracture patterns. Softer weathered material seen between the pillows in places is

Figure 9.13 Deformed volcanogenic rocks and interbedded amphibolite schists, Butterlope Glen.

Figure 9.14 A flat-lying downward-facing fold in quartzite, Butterlope Glen.

considered to be altered **hyaloclastic** material, which is broken pieces of glassy lava formed when the lava was originally in contact with the water. The pillow lavas are close to the limestones of the Dungiven Formation (as well as in Butterlope Glen, there are outcrops of light grey limestone in a small disused quarry pit about 150 m east of the buildings at Craig and also a disused lime kiln nearby). This suggests that the volcanic activity took

Figure 9.15 Pillow lavas of Dalradian age, Craig, County Tyrone.

place within a relatively shallow marine shelf environment. These lavas represent an important volcanic phase within the Dalradian and are significant evidence in unravelling the complex sequence of events within this important period.

Mountfeld Old Quarry Retrace the route from Craig to the B48 and return to Gortin via Plumbridge. From Gortin take the B46 road east for approximately 4 km to Drumlea Bridge (H534858) and take the road to the south. Follow this road for approximately 7 km to a large disused quarry on the east side of the road at H535786 about 500 m from the village of Mountfield. There is limited parking at the entrance to the quarry, which is currently disused. The importance of the old quarry at Mountfield is that it exposes the Omagh Fault, a major southeasterly thrust fault that marks the southern boundary of the Sperrin Mountains Dalradian belt. In the quarry, Dalradian metamorphic rocks overlie the younger Tyrone Volcanic Group, which is Ordovician in age.

At the top of the northwest face of the quarry is a pale green or grey **mylonite** horizon above the dark blue-grey lavas and tuffs of the Tyrone Volcanic Group. "Mylonite" is the term given to broken and smashed up rocks produced by movement on fault surfaces, and this mylonite zone (3 m thick) marks the position of the Omagh Fault (Fig. 9.16). The mylonite consists of a greyish clay matrix containing angular fragments of a mica-rich pelite. These fragments can also be found in the scree below the quarry face. The Omagh Fault is a low-angle southeasterly directed thrust fault that has brought the Dalradian (Mullaghcarn Formation) to rest above the basic lavas and tuffs of the Tyrone Volcanic Group, which is Ordovician in age and therefore younger than the Dalradian. Extreme care must be taken when examining the mylonite outcrop. Access is through the trees along the quarry edge. The outcrop can be extremely slippery, especially in wet conditions, and care should be taken to avoid stepping on it.

Tyrone Plutonic Complex and Tyrone Volcanic Group

Black Rock (H685836) From Mountfield Old Quarry take the A505 Omagh–Cookstown road eastwards approximately 15 km to the brown signpost to the Beaghmore stone circles at the junction at Dunnamore Bridge (H704799). Park in the layby at the stone circles; Black Rock is adjacent to the south. There is an alternative layby at the roadside at H687839 on the east side of the road. Rocks here are part of the Tyrone Plutonic Group, which forms part of the Tyrone Ophiolite complex, which has been interpreted as a fragment of Ordovician island arc, and are broadly comparable with similar rocks at Clew Bay in western Ireland and at Ballantrae on the west coast of Scotland. The main lithology at Black Rock is a coarse

Figure 9.16 Outcrop of the Omagh Fault (dashed line) in Mountfield Quarry with pale green mylonite overlying the dark blue-grey lavas and tuffs of the Tyrone Volcanic Group.

Figure 9.17 Gabbro/dolerite contact at Black Rock, County Tyrone.

hornblende gabbro that varies from being even grained to having very coarse grains of pegmatite gabbro in which some individual hornblende and plagioclase crystals are greater than 2 cm in diameter. The gabbro contains inclusions and xenoliths of an early suite of dolerites, as well as being cut by a series of younger dolerites in the form of 1–2 m-wide dykes (Fig. 9.17). These features are well demonstrated on south-facing crags at H685836. The pegmatite gabbros are exposed on the crags around H684836, to the north of Beaghmore Road.

At Beaghbeg (H674828), about 1 km southwest of Black Rock along the Beaghmore Road, there are extensive outcrops of tuffaceous volcanic rocks and volcanic breccia in the Tyrone Volcanic Group, which may be examined if time permits. From the Black Rock or Beaghmore stone circle proceed along the Beaghmore or Beaghbeg road for access to these crags.

Beaghmore stone circles (H686842; Fig. 9.18) The circles at Beaghmore represent the best-known assemblage of stone circles and rows in Ireland and are well worth visiting. Bog excavation here has uncovered a complex of seven stone circles, several stone alignments and a dozen small stone cairns on a site covering more than 2 ha. The age data from the site indicates occupation of the area from the Neolithic to the Bronze Age, from about 6000 years ago, with the main period of the circles and rows around 3000–4000 years ago. The small stone cairns usually cover a cremation site. There is debate about the extent to which the stone alignments have an astronomical function. It has been suggested, for example, that two of the

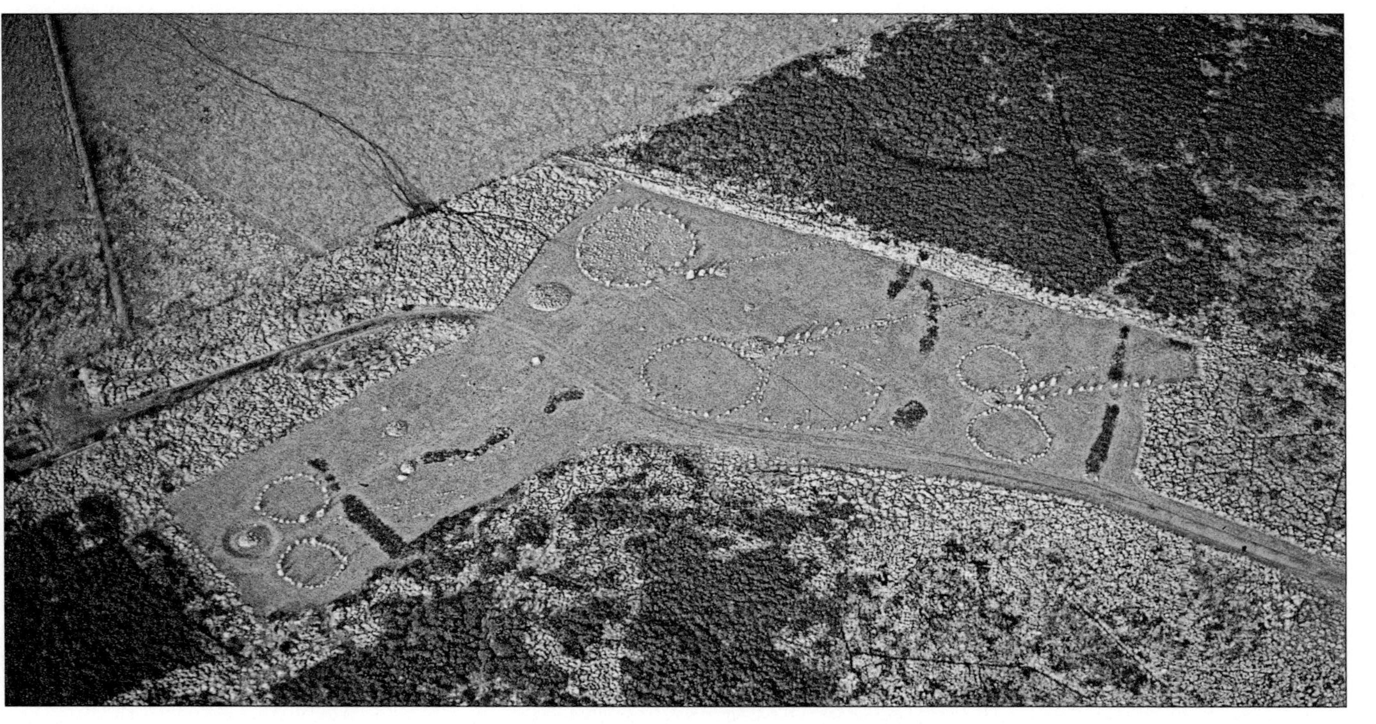

Figure 9.18 Aerial view of the Beaghmore stone circles, County Tyrone. © Crown copyright; reproduced with the permission of the Cortroller of HMSO.

stone alignments at Beaghmore align with the summer solstice, but it seems that the stone walls acted as a site for burial, public rituals and astronomical observations. The organization of these sites suggests a high level of cooperation within the community, comparable with the megalithic monuments seen at Carrowmore and Sligo.

Chapter 10

Down

North Down

The itinerary for north Down covers the northern part of the Ards Peninsula, the Newtownards area, Donaghadee and the Cultra foreshore on the south side of Belfast Lough.* The geology of north County Down is dominated by Ordovician–Silurian sedimentary sequences and associated minor igneous and igneous-derived rocks (pillow lavas, **bentonites, lamprophyre** dykes), Permo-Triassic rocks of Scrabo and Cultra, and the Palaeogene (Tertiary) igneous rocks of the Scrabo area (Fig. 10.1). The marked contrast in scenery between counties Antrim and Down reflects the fundamental differences in the bedrock geology. Antrim is underlain mainly by the relatively young Palaeogene basalts (about 60 million years old) overlying the Cretaceous Chalk (about 100 million years old), the bedrock geology of County Down is dominated by sandstones, shales and mudstones, laid down by turbidity currents in the Ordovician and Silurian periods, 500–400 million years ago. These marine sediments form part of an extensive Lower Palaeozoic belt that was deformed during the Caledonian orogeny and which extends southwest from County Down, through County Armagh to County Longford, and is a continuation in Ireland of the Southern Uplands of Scotland (see Fig. 5.8). They were probably deposited in an ocean trench at the edge of the Iapetus Ocean where the Iapetus oceanic plate was being subducted under the North American continental plate (Fig. 10.2). The diagram shows the position during early Silurian times. This gradual closure of the Iapetus Ocean caused progressively younger slices of ocean-floor sediment to be accreted or plastered onto the side of the continental plate along the northwest margin of the ocean, forming what is known as an accretionary prism.

As mentioned on pp. 36–41, the Iapetus Ocean had developed as the continents Laurentia and Gondwana drifted apart in late Precambrian and Cambrian times. However, by Ordovician times the continents had passed

---

* Sheet 15, 1:50 000, Belfast. GSNI Sheet 37, Newtownards, and sheet 29, Carrickfergus.

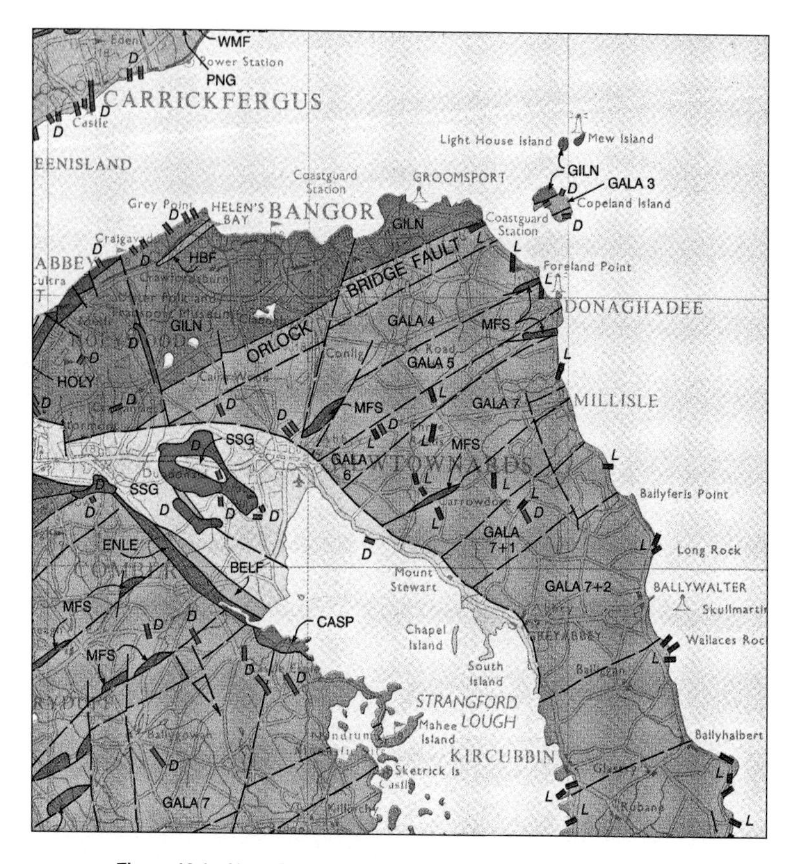

Figure 10.1 North Down and the Ards Peninsula (© Crown copyright).

their maximum separation and the Iapetus Ocean had begun to contract. Ultimately, about 400 million years ago, the continental fragments collided to form the Caledonian Mountains, which extend from Norway through Scotland and Ireland, and into Newfoundland and Nova Scotia in Canada. One effect of these continental closures was to bring together the two parts of Ireland (northwest and southeast), which until then had been separate, and weld them together along a suture line that runs from Clogher Head in County Louth on the east coast, to the Shannon Estuary in the southwest. Across the Irish Sea this suture line continues eastwards along the line of the Solway Firth. The folded and faulted turbidites of north Down are evidence of the compressive forces involved in the closure of the ocean, with the sediments themselves showing evidence of marine sedimentation and associated igneous activity.

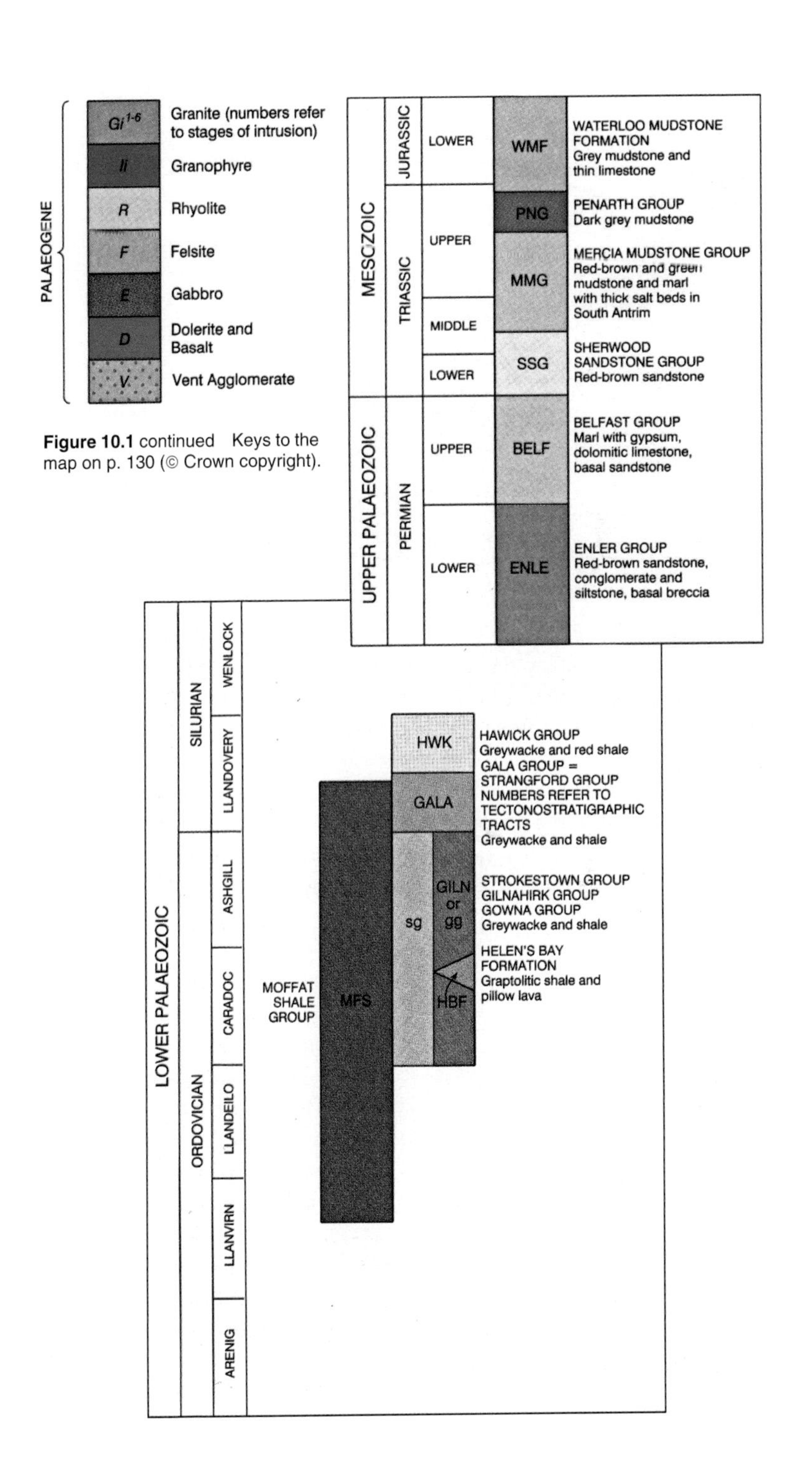

Figure 10.1 continued Keys to the map on p. 130 (© Crown copyright).

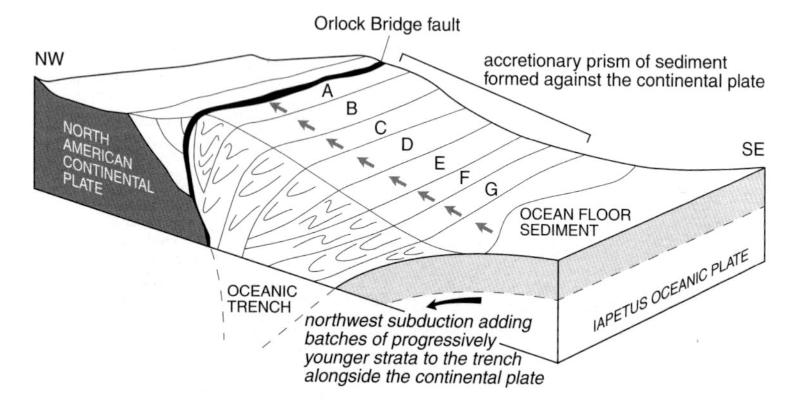

Figure 10.2 Formation of an accretionary prism by subduction of the Iapetus oceanic plate under the North American continental plate during early Silurian times (after Smith et al. 1991). Blocks of sediment become progressively younger from A to G as they are added to the accretionary prism formed against the continental plate with progressive subduction of the Iapetus oceanic plate.

Lower Palaeozoic rocks: Helen's Bay–Donaghadee (Coalpit Bay)–Millisle (Woburn)
This series of exposures, along an approximately 20 km coastal strip, illustrates the main features of the Lower Palaeozoic **greywacke** sequences of the area. Allow 3–4 hours for the full excursion.

Helen's Bay Helen's Bay is located off the A2 Belfast–Bangor road at J455825. Proceed to the car-park at the end of Church Road in the village (J459829). From there the exposure can be reached from the coastal footpath, which runs west from the beach at Helen's Bay. Take the steep path down to the foreshore immediately past the three-storey building with private jetty. The outcrops around the rocky promontory of Horse Rock and the bay area at Ballygrot (Fig. 10.3) expose the oldest sediments in the north Down area. They are Ordovician shales and sandstones of the Gilnahirk Group, including within them shales and pillow lavas of the Helen's Bay Formation. Pillow lavas indicate lava eruption under water and, since the associated sediments here are of marine origin, the interpretation is of volcanic eruption on the ocean floor. Although both sides of the Irish Sea have similar successions of greywackes, mudstones and black shales, and evidence of contemporaneous volcanism is widespread in Scotland, this locality is so far the only example known in County Down.

The rocky promontory is Horse Rock (Fig. 10.3) and close examination of its eastern side towards the seaward end reveals clearly defined pillow structures (Fig. 10.4). The rocks adjacent to the lavas to the northwest are fragmentary and contain recognizable clasts of pillow lavas within a pale clay-rich matrix. Farther to the north, around Ballygrot, are exposures of

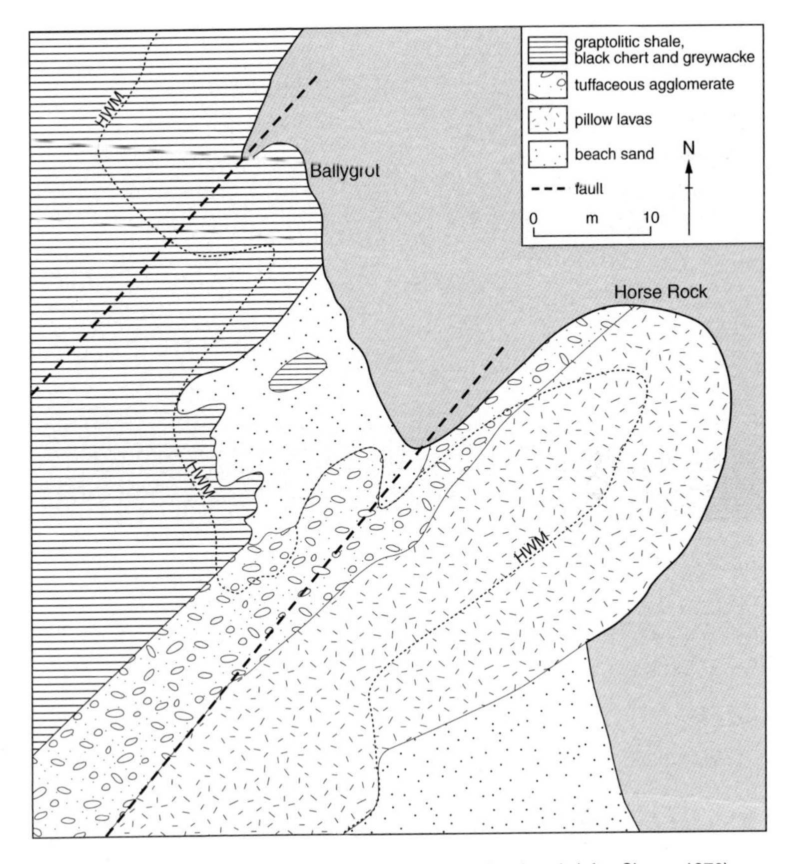

Figure 10.3 Geology of the western end of Helen's Bay beach (after Sharpe 1970).

shale, black chert and greywackes. Fossil **graptolites** in the shale allow the sediments here to be dated as Ordovician (see Coalpit Bay excursion, as follows).

Coalpit Bay, Donaghadee From Helen's Bay return to the A2 road and proceed east to Donaghadee (J590800). Take the A2 through Donaghadee towards Millisle to Coalpit Bay at J596788. Access to the foreshore here is via the recreation grounds beside the car-park and toilets. On the foreshore here at Coalpit Bay are several interesting exposures that clearly illustrate the complex faulting and folding of the Lower Palaeozoic sediments. This faulting has resulted in the occurrence of an isolated block of older Ordovician rocks within younger Silurian rocks, a geological phenomenon known as an inlier (Fig. 10.5). The section also demonstrates the uses that can be made of fossils, in this case floating or planktonic animals

Figure 10.4 Pillow lavas at Horse Rock, Helen's Bay.

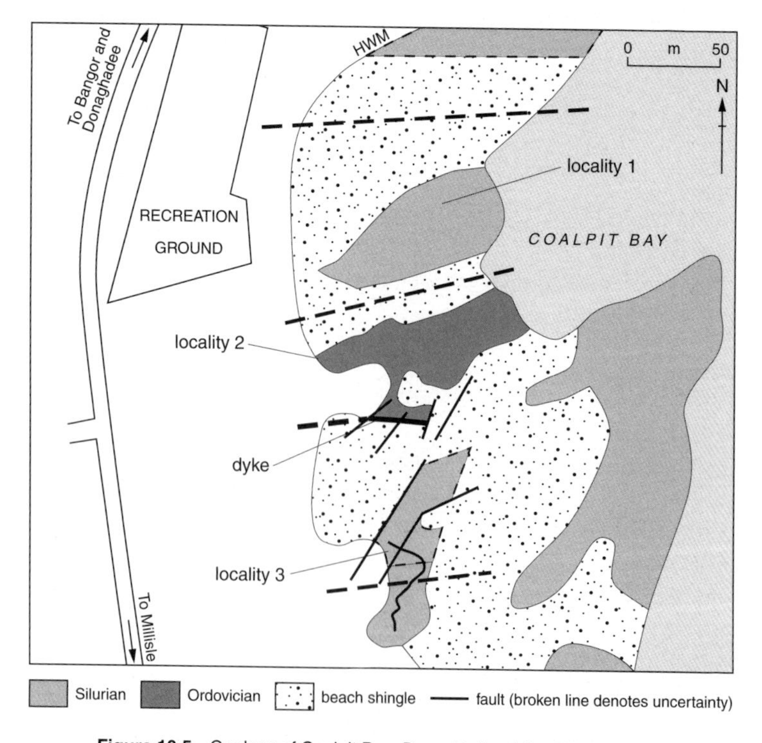

Figure 10.5 Geology of Coalpit Bay, Donaghadee (after Wilson 1972).

known as graptolites, in unravelling the stratigraphy of such complicated successions. There are also further examples of igneous activity in the area, contemporaneous with both the sediment deposition and later events.

Those outcrops near the southern end of the tennis courts (locality 1), and below the high-water mark, are folded into parallel anticlines and synclines with the axes trending roughly east–west. These rocks are Silurian. Farther south across a stretch of shingle beach, the rocks in the next set of exposures (locality 2) are much more broken up and shattered, with fewer thick sandstone bands. Fossils (graptolites) show them to be Ordovician and the stretches of shingle beach directly north and south of the outcrop probably represent the faulted boundaries (Fig 10.5). Also, on the southern side of the outcrop is a well exposed dyke, trending roughly east–west and offset by a small fault at its seaward end. This dyke is just one of a group, all with similar composition and orientation, together known as a **swarm**. They postdate most of the folding and faulting of the sediments. They are intermediate in composition between basalt and granite, termed lamprophyre, and some of them are related to the intrusion of the Newry granodiorite complex, which lies about 30 km to the southwest.

South from the landward edge of the dyke outcrop (locality 3) are thin sandstones with interbedded dark mudstone bands and several pale clay layers (Fig. 10.6). Pale grey or grey-green in colour, these are known as bentonites and are composed of clay minerals (mostly the variety montmorillonite) formed from weathering of volcanic ash that fell into the sea. The base of each bentonite is sharply defined, but some beds grade

Figure 10.6 Mudstones and interbedded bentonite clays formed from weathered volcanic ash, Coalpit Bay, Donaghadee.

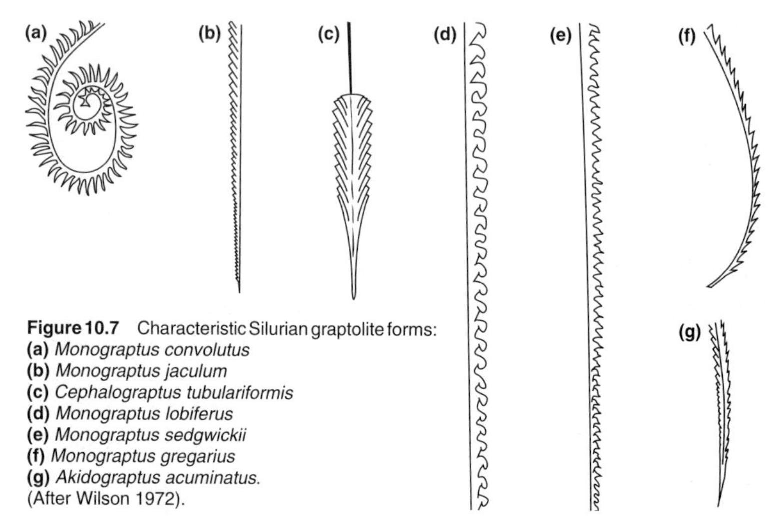

Figure 10.7 Characteristic Silurian graptolite forms:
(a) *Monograptus convolutus*
(b) *Monograptus jaculum*
(c) *Cephalograptus tubulariformis*
(d) *Monograptus lobiferus*
(e) *Monograptus sedgwickii*
(f) *Monograptus gregarius*
(g) *Akidograptus acuminatus*.
(After Wilson 1972).

upwards into darker mudstone. The bentonites provide further evidence (see also Helen's Bay) of contemporaneous volcanic activity in or near the environment of deposition of the turbidites, perhaps an island-arc type of plate margin. The dark mudstone bands contain abundant graptolites, for example *Monograptus convolutus* and *Monograptus gregarius* (Fig. 10.7), which confirm a Silurian age. The graptolites now occur as pencil-like marks on the bedding planes of the mudstone and can often be found by careful splitting of the darkest layers with a fine hammer or chisel.

Galloway's Burn (J595783) Along the coast some 300 m south of Coalpit Bay is a wooden bridge across a narrow stream, Galloway's Burn. About 200 m farther south from this bridge, near the high-water mark and beside a new house on the point before the next bay to the south, are superb examples of what are known in sedimentology as **sole structures** (Fig. 10.8). These are features that form on the sole or bottom of a sandstone or mudstone bed, particularly in environments where turbidity currents are common. As a turbidity current passes over a bed of mud on the ocean floor (see Fig. 5.7), turbulent eddies at the head of the current scoop out a spoon-shape depression in the mud, with the deepest part of the depression up current. As the tail of the current passes over the depression, it is filled with fine-grain sand. This sand is then preserved as a downward projection from the base of the sandstone. In the examples here, the base of the sandstone is exposed because the succession has been overturned and the original underlying mudstone eroded off. As well as providing evidence of turbidity currents, sole structures also provide proof that a succession has been overturned – so-called "way-up evidence". When a

Figure 10.8 Sole structures, Galloway's Burn, Donaghadee.

series of examples were measured around Galloway's Burn, the currents trended almost due west in this area.

Woburn Foreshore, Millisle (J613745) From Coalpit Bay proceed east on the main road to Millisle. Approximately 2 km south of Millisle on the road to Ballywalter, park at the picnic site on the seaward side of the road just past Woburn House. Access to the foreshore is by the southern end of the car-park. At this point and near the high-water mark is a conspicuous syncline in the Silurian Strangford Group rocks (Fig. 10.9), providing clear

Figure 10.9 A syncline in Silurian sediments at Woburn, Millisle.

evidence on a relatively small scale of the nature and extent of the folding that affected these Lower Palaeozoic sediments.

Permian and Triassic rocks

By about 250 million years ago a single continent, Pangaea, had developed, comprising all the coalesced continental fragments, including the northern continent Laurentia and the southern continent Gondwanaland. The change in continental distribution on the Earth's surface resulted in radical changes in climate and the succeeding Permian and Triassic periods are dominated, like the earlier Devonian period, by semi-arid desert conditions. Since Ireland lay deep within this supercontinent and only a few degrees north of the Equator, it was a hot desert with a covering of angular weathered rock fragments that consolidated to form a rock known as breccia. Outcrops of Permian rocks are rare in Ireland, but a good exposure can be examined on the foreshore, at Cultra, on the south side of Belfast Lough. During the Permian, Ireland was briefly invaded by a shallow sea from the north, which deposited a magnesium-rich limestone in the Cultra area. A major extinction event marked the end of the Palaeozoic era and the first period of the Mesozoic era (the Triassic), represented in this area by the Triassic sandstones of the Scrabo area, near Newtownards. By this time Ireland had drifted north of the Equator to around the latitudes of the present Sahara, but still remained part of Pangaea. Allow approximately 3 hours for the Cultra and Scrabo localities.

Permian rocks at Cultra, Belfast Lough foreshore

From Millisle proceed via the A2 and Bangor towards Holywood. Approximately 2 km east of Holywood is the Ulster Folk Museum on the south side of the road and the Culloden Hotel on the north side. Just west of the Culloden Hotel is Cultra Avenue, leading to the foreshore at Cultra (pronounced "Cultraw") at J412809. Follow this road past the Royal North of Ireland Yacht Club and proceed along Sea Front Road to the end of the public road and park. Access to the beach is via a set of steps adjacent to where the public road swings inland again to the right.

Rocks of Permian age outcrop in only two localities in the north of Ireland, one of which is here at Cultra. The Permian period in Ireland marked a return to mainly hot desert conditions, which replaced the marine or deltaic conditions of the preceding Carboniferous. Across most of the north of Ireland, the early Permian was mainly a period of subaerial erosion, with eroded debris from the high ground accumulating in screes along the edges of mountain ranges, forming deposits of angular rock fragments

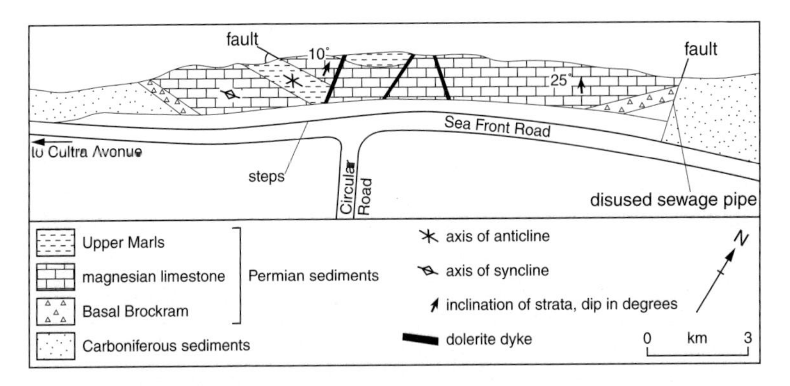

Figure 10.10 Geology of the Cultra foreshore (after GSNI 1982).

known as breccias. The Upper Permian beds were deposited in what has become known as the Bakevellia Sea, which advanced from the north and periodically covered eastern Ireland, depositing a magnesium-rich limestone, marls and mudstones. The succession exposed at Cultra is:

Permian Upper Marls
Magnesian Limestone
Breccia (Brockram)
unconformity
Lower Carboniferous

The Permian rocks are best exposed on the section of beach between the steps and the pipeline, from the now fortunately disused Cultra sewage works, that crosses the beach some 150 m to the east (Fig. 10.10). The lowest Permian unit, the Breccia or Brockram, is about 2 m thick and comprises mostly angular clasts of weathered Lower Palaeozoic greywackes, unconformably overlying the Carboniferous mudstones (Fig. 10.11). The Brockram is, in turn, overlain by about 8 m of massive yellow dolomite, the Magnesian Limestone. Dolomite is a form of limestone containing the mineral dolomite (calcium magnesium carbonate: $CaMg(CO_3)_2$) rather than the mineral calcite (calcium carbonate: $CaCO_3$). There are occasional thin shelly bands within the Magnesian Limestone; the contact with the overlying Upper Marls is well exposed and sharp (Fig. 10.12). The foreshore sediments, both Carboniferous and Permian, are cut by Palaeogene dykes, one of which has been almost completely altered to soft yellowish clay material. The sewage outfall pipe to the east marks the line of a northwest-trending fault that brings Permian rocks down against the Carboniferous rocks to the east. The line of the fault shows brecciation and disruption of the sediments. At the western edge of the Permian outcrop near the access steps there is a small syncline exposed in the Upper Marls

Figure 10.11 Junction of Magnesian Limestone over basal Permian Brockram, Cultra foreshore.

with a parallel anticline in the Magnesian Limestone to the west. Below this a thin bed of Brockram lies unconformably on the Carboniferous sediments.

Triassic rocks of the Scrabo area

From Cultra retrace the route back to main road A2 and proceed west in the direction of Belfast. Turn south up the hill at the first road on the left (Whinney Hill) and follow the signs to Newtownards. In Newtownards follow the signs to the Scrabo Country Park and the car-park at J476722.

In Triassic times the area around Newtownards was a lowland valley or trough, being steadily infilled by terrestrial sediments. Since the

Figure 10.12 Junction of the Upper Marls over Magnesian limestone, Cultra foreshore.

climatic environment was a semi-arid desert with periodic torrential rainstorms, these sediments are mainly of sandstone and mudstone deposited in the ephemeral lakes formed after intermittent periods of heavy rain. Within these sediments, ripple marks provide evidence for shallow-water conditions, and mud cracks and reptile footprints for periodic drying out. In the nineteenth century the pale sandstones of the Newtownards area, particularly those seen in the extensive former quarries below Scrabo Hill, were used widely as a building stone in the nearby town of Newtownards and as far away as Belfast and Dublin.

The sandstones in the Scrabo quarries were weakly metamorphosed as a result of igneous activity during Palaeogene times. The resultant slight hardening has produced a more durable stone than would otherwise have been the case.

The sandstones are part of the Sherwood Sandstone Group, referred to in earlier descriptions as the Bunter or New Red sandstone. They are fine- to medium-grain sandstones with subordinate dark-brown mudstone and siltstones interbedded. Cross-bedded sandstones are common; these are inclined beds in a sedimentary rock, formed by currents of wind or water in the direction in which the bed slopes downwards. Scrabo today forms a prominent hill, some 160 m high, to the southwest of Newtownards.

The monument Scrabo Tower was erected to the Marquess of Londonderry, in the late 1850s. Within the tower is an exhibition of the ecology, geology and archaeology of the County Park, which is generally open from Easter until the end of the summer. The hill forms a crag-and-tail feature, with the tail extending southeastwards, demonstrating ice movement from the northwest. The prominence of the hill is attributable to the capping of the relatively soft Triassic sandstones by the resistant Palaeogene dolerite sill on which the tower stands.

South Quarry From the car-park take the track leading to the tower and, after 100 m, turn down hill towards the South Quarry (Fig. 10.13). The main face of the quarry can be viewed from this path, but no attempt should be made to climb over the fence. Here a thick succession of sandstones beneath the main Scrabo sill shows cross bedding at various scales, with the sandstones intruded by sills and a dyke of Palaeogene age (Fig. 10.14).

Along the path is an Environment and Heritage Service information board detailing various aspects of the geology and wildlife of the area, including peregrine falcons and ravens. In the western bay of the quarry an agglomerate-filled vent is visible, with blocks of basalt and sandstone in a soft greenish matrix. Some brecciation can be seen in the quarry wall near the vent, where gas explosions have disrupted the sandstone beds and stained them a dark purple colour. Near the entrance to the South

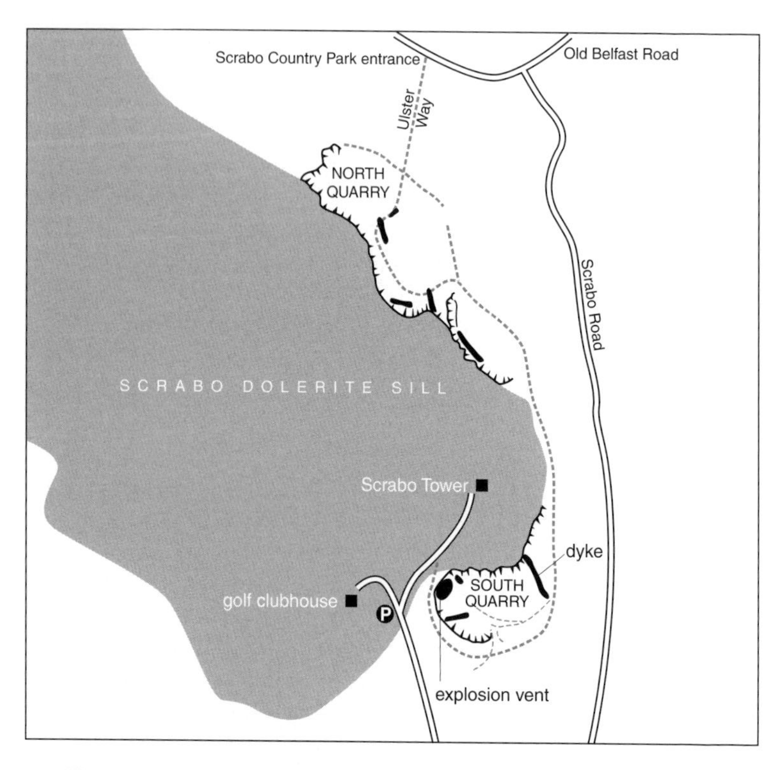

Figure 10.13 The main geological features around Scrabo Tower, Newtownards.

Figure 10.14 Dolerite sills (Palaeogene) intruding Triassic sandstones, south quarry, Scrabo Hill, Newtownards.

Quarry is a conspicuous dolerite dyke, which has been left as a pinnacle after quarrying, and which can be seen to cut through, and therefore post-date, some of the smaller sills. The rotted base of the main sill that caps the hill can be seen in the upper part of the South Quarry face above several smaller dolerite sills that are also exposed in the quarry face. These smaller sills can be seen to change level within the sandstones they are intruding, a process known as transgressing.

The path through the Country Park is part of the Ulster Way and it passes through the Middle and North Quarries, showing further examples of sedimentary structures within the sandstones and the remains of tramways that were installed in the quarries by the 1860s.

*The Mourne Mountains and south Down**

Introduction

The landscape of the southern half of County Down is dominated by the granitic Mountains of Mourne, with Slieve Donard at 852 m the highest peak. Along with the mountains around Carlingford and Slieve Gullion in counties Armagh and Lough to the southwest, they represent the eroded remains of the final stages of the igneous activity that occurred in north-east Ireland during the formation of the North Atlantic. The activity at Carlingford and Slieve Gullion formed volcanic peaks with lava flows at the surface, as well as complex intrusive features, but there is no evidence that the magma from the Mournes ever broke the surface of the crust. Instead, a series of intrusive phases from two separate centres produced a complex of five distinct granite types, named G1–G5, across the approximately 20 km width of the intrusion. The internal contacts of the five granite varieties are shown on the GSNI Mourne Mountains Special Sheet, 1:50 000 Series. The granite localities examined here are G2, which is a quartz-rich biotite granite with abundant dark quartz. This excursion explores the relationship between the Palaeogene (Tertiary) granites and the much older Silurian sediments into which they are intruded (Fig. 10.15). These sediments are the sandstones and mudstones that underlie most of counties Down and Armagh. Where the granite has intruded them, they have been baked or thermally altered to form a hard and often striped rock – hornfels. In addition to the main intrusive body of the granite pluton, there is an extensive suite of minor intrusions associated with the activity. As well as linear dykes there are also **cone sheets**, which are roughly funnel-shape dykes, often surrounding igneous intrusions.

---

* GSNI: *Mourne Mountains*, special sheet; OS sheet 29, *The Mournes* – 1:50 000.

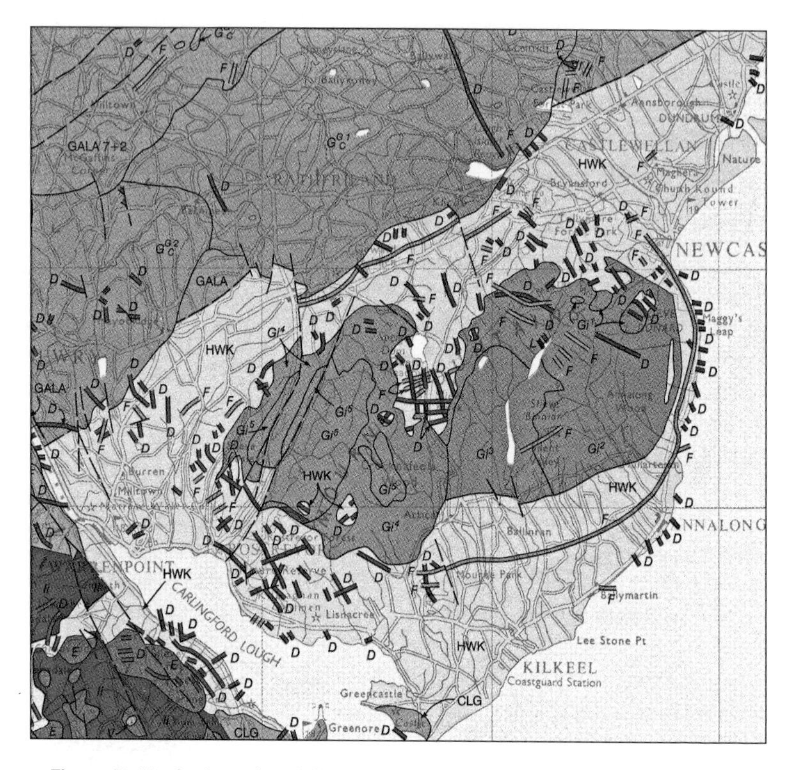

Figure 10.15 Geology of south Down. See pp. 145–146 for keys. (© Crown copyright).

Eastern Mournes

The first three sections of this part of the excursion are free standing and can be followed separately. Proceed from Newcastle south along the A2 to the car-park at Bloody Bridge (J389269).

Coastal dyke swarm – Bloody Bridge to Green Harbour

Along this stretch of coast, Silurian greywackes are intruded by dykes of basalt, andesite and granite composition, often as multiple and composite intrusions. Multiple dykes occur when the same fissure is used more than once to intrude magma of the same composition. Composite intrusions are those multiple intrusions where the magma composition changes. At Green Harbour (J388250) is a composite dyke 8 m wide, with basaltic margins and a granitic interior zone. This is on the south side of the stream about 50 m south from the old winch, near the high-water mark. A study by Tomkeieff & Marshall (1935) recorded details of more than a hundred dykes along this section of coast from Newcastle to south of Annalong. Return to Bloody Bridge car-park; allow an hour for the return walk.

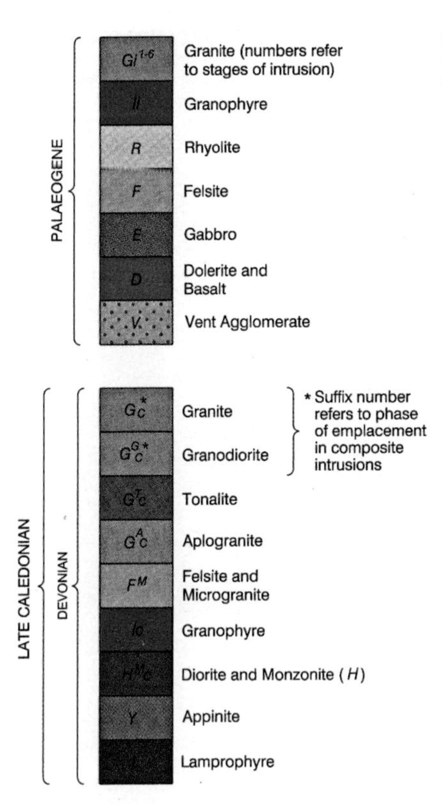

Figure 10.15 continued Part of the key for the map on p. 144 (© Crown copyright).

Bloody Bridge River section
Leave from the Bloody Bridge car-park at the south end, cross the road and follow the track on the north side of the Bloody Bridge River. Hornfelsed sediments exposed in the river become increasingly striped as the contact with the granite is approached. The greenish coloration is attributable to the mineral pyroxene (the variety diopside) and the blue-black bands are rich in dark mica (biotite). About 600 m up stream from the bridge, G2 granite is in near-vertical contact with hornfelsed sediments. The granite near the contact is of fine to medium grain. A further 200 m up stream, a double-pipe aqueduct crosses the river. Up stream from the pipes there is a further 100 m-wide section of hornfelsed sediment that then gives way to the main body of the G2 intrusion (Fig. 10.15). The contact is clearly visible on the river's north bank, where it flows down a steep-sided gorge. Return to the car-park via the track; allow an hour for the round trip.

Glassdrumman Port (J380221) At Glassdrumman Port, about 5 km south of Bloody Bridge, is a superb example of a composite cone sheet, well

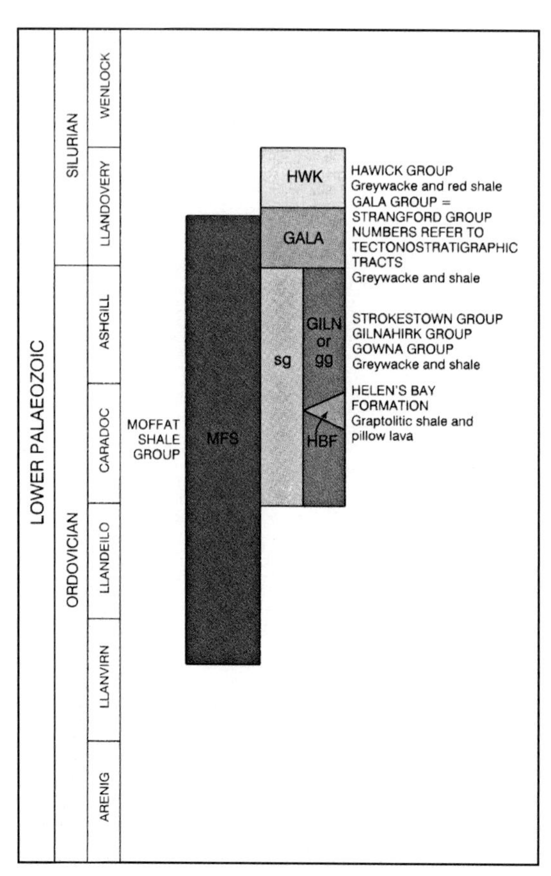

Figure 10.15 continued Part of the key for the map on p. 144 (© Crown copyright).

exposed in contact with steeply dipping Silurian sediments. Access is via the lane approximately 300 m south of the Roman Catholic church at Glassdrumman village. One of the gate pillars at the entrance of the lane has a post box; there is limited parking for cars only. The main road here is narrow and busy, and care must be taken to avoid obstructing the traffic. Proceed down the lane to the coast and go north about 200 m. Across a small stream there are excellent exposures of a thick cone sheet, approximately 15 m wide, with basaltic margins and a microgranite interior (Fig. 10.16). The texture of the microgranite is referred to as **porphyritic**: it contains relatively large crystals called **phenocrysts** in a finer matrix. In this case the phenocrysts are quartz and feldspar, and can be easily seen. The microgranite is sometimes referred to as quartz–feldspar porphyry (Figs 10.16, 10.17). The inner and outer contacts of the basaltic–granitic components of the cone sheet are well exposed and show some degree of mixing or hybridization of the two magma types (Fig. 10.18). Also, some of the

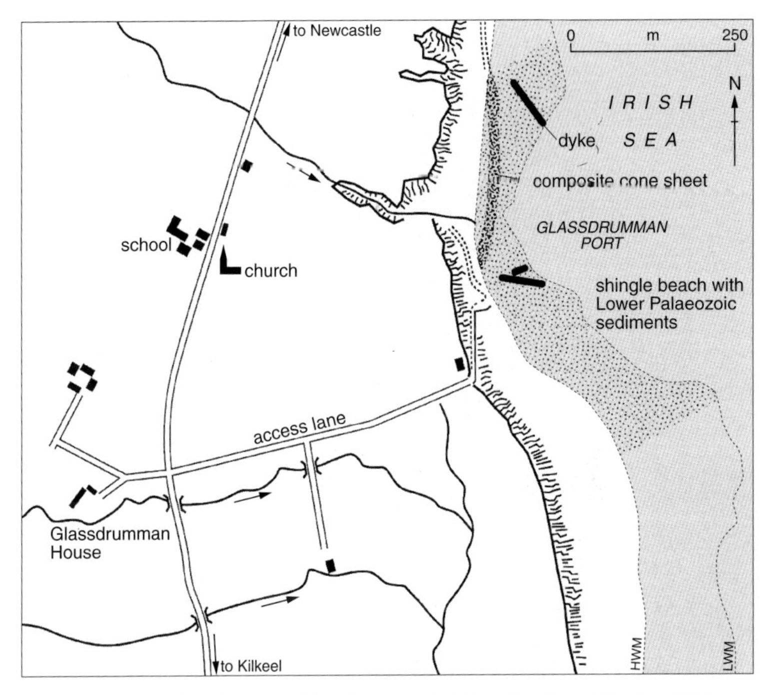

Figure 10.16 Sketch map of Glassdrumman Port (from Emeleus & Preston 1969).

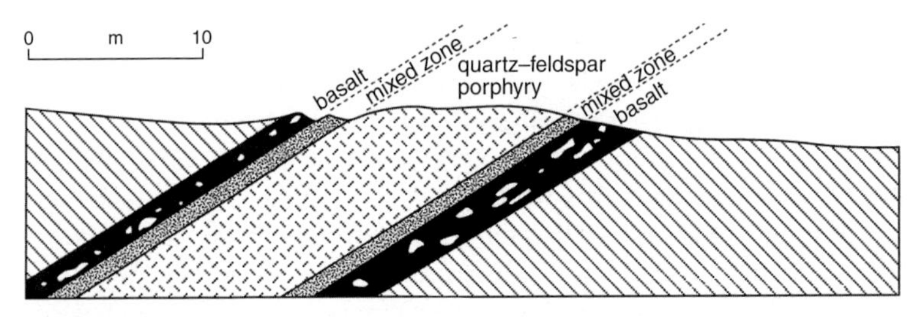

Figure 10.17 Cross section through the Glassdrumman composite cone sheet (from Tomkeieff & Marshall 1935).

large feldspar phenocrysts from the microgranite are incorporated within the basic margins of the cone sheet. The Silurian sediments here dip south at around 40° and the basaltic margins are chilled against these country rocks. Return to vehicle via the lane; allow an hour.

Northern Mournes

This excursion encompasses the Glen River section and examines the wall and roof relationships between the granite and the intruded sediments, with mineral collecting at the Diamond Rocks. It can take half a day, dealing

Figure 10.18 Glassdrumman cone sheet, showing the contact between the basic component and the central granitic component on the inner margin.

only with the Glen River section and the wall and roof relationships, or a whole day to include the Brandy Pad and the Diamond Rocks. It should be noted that this second option requires a full day among some of the highest and most remote areas of the eastern Mournes, where rapid weather changes can render conditions hazardous at any time of the year, and appropriate footwear and waterproof clothing are essential.

Glen River section Proceed to the car-park at Donard Park (J372305) on the south side of Newcastle. Leave the car-park at the southern end and follow the path up the Glen River. At the first bridge there are steeply dipping banded hornfelses, with dark bands rich in mica (biotite) and greenish bands rich in pyroxene (diopside). Continue upwards to the second bridge (J369299) where the path crosses to the north side of the stream. The hornfels/granite contact can be seen in the river about 10 m up stream from a prominent waterfall. The contact almost coincides with a Palaeogene dolerite dyke, about 200 m down stream from the third bridge at J366297 (Fig. 10.19). The track emerges from the forest just up stream from this third bridge. There is a refurbished icehouse on the southern side of the river 100 m from the forest boundary (prior to the development of refrigeration the icehouse was used to store winter ice in cool conditions for summer use). Follow the track upwards to the edge of the forest on the north side of the track at J357291. Turn northwest and follow the fence up slope to where it turns northeast along the boundary of the forest. Approximately 200 m along this fence from the corner are outcrops of near-horizontally bedded hornfels in contact with granite G2. This exposure represents part of the roof of the intrusion (Fig. 10.20).

Figure 10.19 Granite/hornfels contact, Glen River section, Newcastle.

Follow the contours around the slopes of Slievenamaddy to J360295, where the near-horizontal hornfels beds are clearly seen to overlie the granite, with the contact at a low angle. If the excursion is taken as a half day, then retrace the path back to the car-park at Donard park. Allow three hours for the round trip. If not, then proceed to the instructions for the Brandy Pad and the Diamond Rocks.

Brandy Pad and Diamond Rocks Return to the Glen River track and continue upwards to the saddle between Slieve Donard and Slieve Commedagh and the boundary wall that marks the catchment area of the Belfast Water Commissioners. This massive dry-stone wall was built for the Belfast Water Commissioners in the early part of the twentieth century to mark the extent of their catchment area for the water supply to Belfast. It is about 1 m wide, often as much as 2 m high, and runs for a distance of more than 30 km. Cross the wall and pick up the track heading northwest.

Figure 10.20 Near-horizontally bedded hornfels in contact with the roof of granite intrusion G2.

This is the Brandy Pad and it runs from the Bloody Bridge valley around the western slopes of Slieve Donard and Slieve Commedagh to the Hare's Gap and the Trassey River valley. The Brandy Pad ("path") was named after the smuggling activities in the eighteenth and nineteenth centuries along the coast of south Down. The contraband goods were landed on the shore, often on open beaches, and then carried through the mountains and distributed in the countryside north of the mountains. The valley to the south of Slieve Commedagh is the glacially modified Annalong Valley. Approximately 2 km along the Brandy Pad at J327287 are the Diamond Rocks. This is a **drusy** phase of G2; that is, the rock has many gas cavities in which have developed crystals of smoky quartz, feldspar, mica, beryl and topaz. The locality is found about 60 m above a large boulder of drusy granite, just above the Brandy Pad (Fig. 10.21), about 400 m from the col at the Hare's Gap. Abundant loose blocks yield good specimens and there is little need for hammering. Continue on the Brandy Pad to the Hare's Gap and follow the Trassey River down to the car-park near Trassey Bridge at J312314, if a pick-up has been arranged (Fig. 10.22). Allow for a further walk of about an hour. Alternatively, retrace the path back along the Brandy Pad and back down the Glen River to Donard Park. In this latter case, allow 6–7 hours for the round trip from Donard Park to the Diamond Rocks and return.

Figure 10.21 Large boulder of drusy granite above the Brandy Pad and just below the Diamond Rocks locality, about 400 m from the Hare's Gap.

Figure 10.22 A view of the Brandy Pad looking towards the Hare's Gap (centre field) and Slieve Bernagh to the west.

Chapter 11

Antrim

The Causeway Coast

Portrush to Ballintoy

The Causeway Coast of north Antrim contains some of the most spectacular geology to be seen anywhere in Ireland, and in the Giant's Causeway has the most popular visitor attraction in the country.* The area from Portrush to Ballycastle has been designated an Area of Outstanding Natural Beauty (AONB) by the Department of the Environment (NI) and the Causeway itself has been accorded the status of a World Heritage site because of its importance in the development of geology as a science.

The area includes not just the columnar basalts associated with the Giant's Causeway but a wide range of rock types and geological phenomena, well exposed on the coast and in the cliff sections.

Introduction

The geology of the area is shown in Figure 11.1. Between Portrush (C856410) and Ballintoy Harbour (D040455) the oldest rocks exposed are the Waterloo Mudstone Formation (the Lias Clay), a subdivision of the Jurassic period (190–135 million years ago) and found at White Park Bay (D020440). These generally soft and fossiliferous marine clays occur directly below the chalk of the Ulster White Limestone Formation, which underlies the Palaeogene basalt lavas over much of County Antrim. The chalk is Cretaceous in age (135–65 million years ago) and represents deposits of microfossils on the floor of a warm sea that covered most of Europe as far east as the Urals in Russia. After the retreat of this sea, the resultant undulating land surface was buried by the successive lava flows of the Palaeogene Antrim Lava Group (about 60 million years ago) in an

---

* OS maps: 1:50 000 sheet 4, Coleraine, sheet 5, Ballycastle; Geological Survey of Northern Ireland: sheet 7, The Causeway Coast; *Geology of the Causeway Coast*, H. E. Wilson & P. I. Manning (1978; Memoir of the Geological Survey of Northern Ireland; 2 volumes). *A geological excursion guide to the Causeway Coast*, P. Lyle (1996; Environment and Heritage Service, Department of the Environment (NI)).

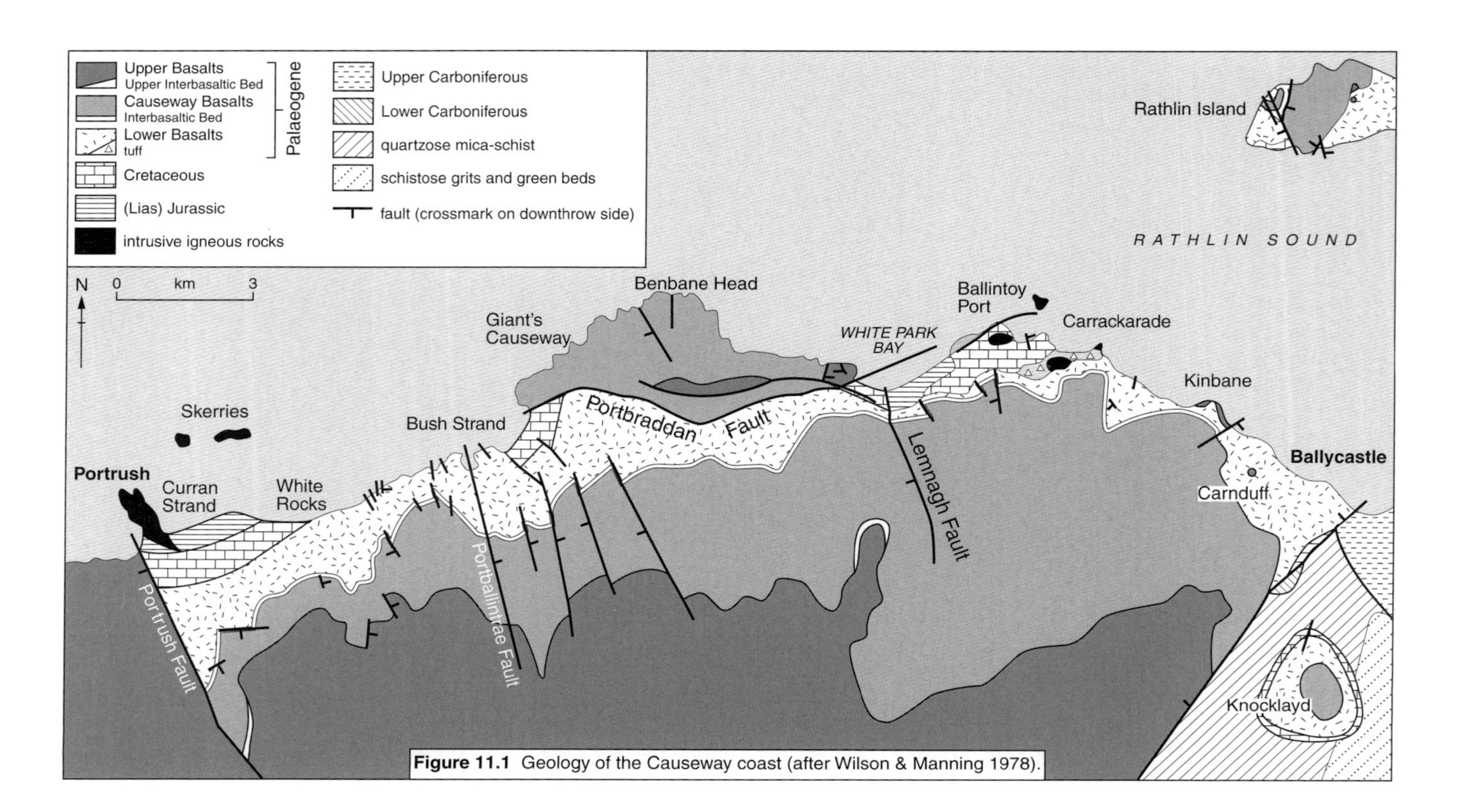

Figure 11.1 Geology of the Causeway coast (after Wilson & Manning 1978).

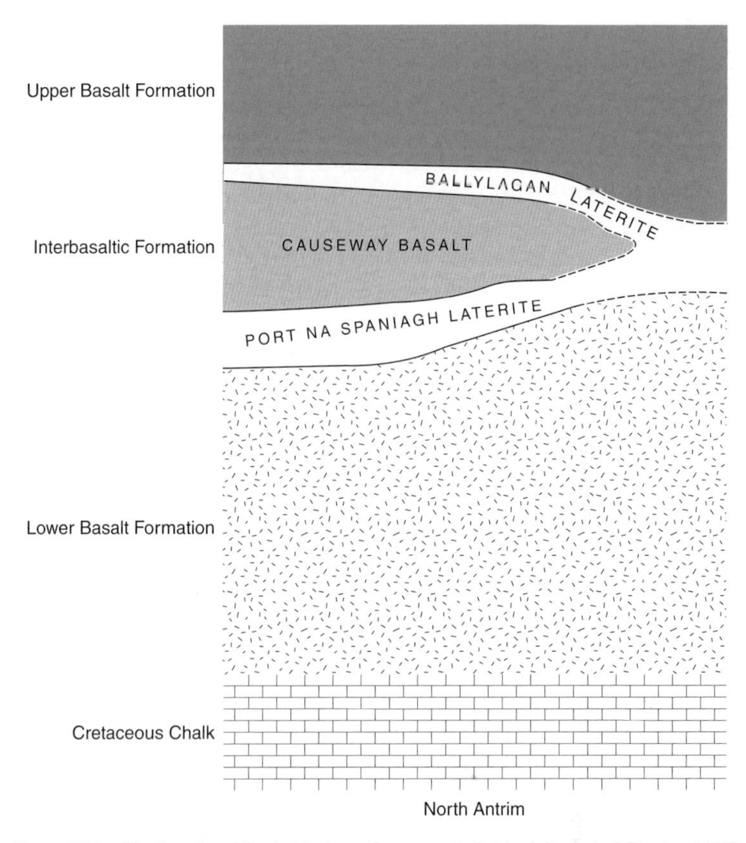

Figure 11.2 Stratigraphy of the Antrim Lava Group, north Antrim (after Lyle & Preston 1993).

eruptive phase associated with the continental rifting that separated Europe from North America. It is this volcanic activity that remains the predominant geological feature of the Causeway Coast. The sequence of lava eruptions is summarized in Figure 11.2. The first phase of eruption in Antrim produced the Lower Basalt Formation (LBF); flows are generally 5–10 m thick, each flow top marked by a purple amygdaloidal zone, often separated by a reddened volcanic dust layer from the flow above. Following a long pause in eruption, the top of this LBF sequence was deeply weathered to produce the Interbasaltic Formation of bright red laterite, many metres thick in places, which separates the LBF from the succeeding Upper Basalt Formation (UBF) over much of central and south Antrim. In north Antrim, however, there was a renewal of activity in Interbasaltic times to produce the basalts of Giant's Causeway type. A further pause in eruption led to renewed lateritization of the top of the Causeway lavas, before eruptions started again to produce the final sequence of flows, the

Upper Basalts. The Causeway lavas are very different from those of the Lower Basalts and are characterized by their columnar jointing, often remarkably regular, as typified by the Giant's Causeway.

The Portrush sill Proceed to Portrush and follow the signs past the harbour to the recreation grounds and the Nature Reserve and Countryside Centre at Lansdowne Crescent (C857414). The Portrush sill is of particular interest for several reasons. It is the largest outcrop of intrusive rock in the AONB, forming the prominent headland of Ramore Head underlying much of the town of Portrush to the south. It continues off shore in the chain of islands, the Skerries, and is an outcrop of major historical significance.

The sill is dolerite, and exposures near the Blue Pool and the Countryside Centre show the contact between the roof of the sill and the surrounding country rock of Lias mudstones (Waterloo Mudstone Formation). The geology of the area is shown in Figure 11.3. These normally soft rocks have been baked by the heat of the intrusion into a hard fine-grain brittle rock

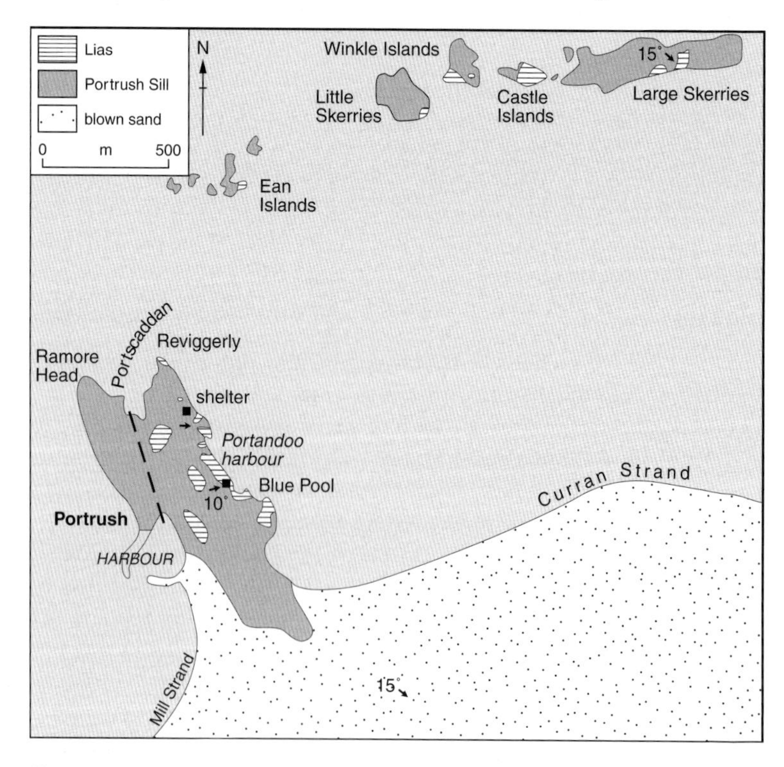

Figure 11.3 The geology of the area around the Portrush sill (after Wilson & Manning 1978). The approximate direction and amount of dip are shown by the larger patches of Lias shale immediately overlying the dolerite of the Portrush sill.

Figure 11.4 Ammonite fossils in baked sediments in contact with the Portrush sill. © Crown copyright; reproduced with the permission of the Controller of HMSO.

known as hornfels. These baked sediments, known historically as Portrush Rock, contain abundant fossils, particularly ammonites (Fig. 11.4) and they featured in a bitter scientific controversy between the Neptunists, who believed that all rocks crystallized from sea water, and the Vulcanists, who favoured a volcanic origin for those rocks they recognized as crystalline. The appearance of fossils in what was erroneously thought to be an igneous rock led Reverend William Richardson in the late eighteenth century to use this occurrence in support of the Neptunist case. In fact he had failed to recognize the contact between these dark fine-grain altered sediments and the underlying coarse dolerite of the sill. By the early years of the nineteenth century this mistake had been acknowledged and the metamorphic or changed nature of the mudstones had been recognized. By then the true origin of igneous rocks had been recognized, the Vulcanists had triumphed, and Neptunism had been confined to history. The contact shown in Figure 11.5 is near the high-water mark close to the navigation triangle and midway between the shelter and the Countryside Centre. The hornfels tends to weather to a colour lighter than the dark-brown weathered dolerite. The near-horizontal bedding planes of the hornfels show good examples of ammonite fossils, and, since this is a conservation area, no attempt should be made to collect specimens by hammering. The dolerite in the sill near the contact is of relatively fine grain because of chilling against the mudstone, becoming coarser towards the interior where cooling was slower. This is readily demonstrated on the dolerite exposures on Ramore Head.

Figure 11.5 Contact between dolerite and hornfels, Portrush sill.

The basalts of the Giant's Causeway From Portrush take the A2 coast road to Bushmills (site of the world's oldest licensed whiskey distillery and generally open to visitors) and from there follow the signs to the visitor centre at the Giant's Causeway. Car-parking and other facilities are available here. The main features to be examined in this part of the excursion are the columnar basalts of the Giant's Causeway, the laterites of the Interbasaltic Formation and the Lower Basalts. The lower cliff path is now closed for safety reasons from just east of the Causeway and no attempt should be made to pass the gate and barrier. Visitors are reminded that the Causeway is a World Heritage site; as such it is a no-hammer zone and the collection of samples is prohibited. From the visitor centre proceed down the path towards the Causeway.

The lava forming the Giant's Causeway flowed into a wide valley, eroded in the Lower Basalt surface before eruption of the Causeway basalts, and in places formed a pond of lava about 100 m thick. The western wall of this valley can be seen beside the path about 200 m from the visitor centre. Here an outcrop of bright red laterite, the Port na Spaniagh laterite (see Fig. 11.2) marks the top of the Lower Basalts, with the first flow of the Causeway basalts lying above it (Fig. 11.6). Seawards the flows of the Lower Basalts are exposed in the cliffs below the Causeway Hotel, and on the foreshore is the sea stack known as the Camel's Back, which is formed by a dyke cutting through the Lower Basalts. Farther down the path towards the Causeway, just before it turns to the east, is a very fine example of Lower Basalts showing **spheroidal weathering**. Here, large blocks

Figure 11.6 Port na Spaniagh laterite underlying the first flow of the Causeway basalts, near the Giant's Causeway visitor centre.

of basalt have been rotted by physical and chemical weathering to form rounded boulders with concentric layering. It is sometimes referred to as onion weathering, as the layers have the appearance of the internal structure of an onion. From here there are clear views of the Causeway to the east, with the Interbasaltic laterites of the other side of the palaeo-valley exposed on the east side of Port Noffer (Figs 11.7, 11.8). The Lower Basalt lavas can be examined on the eastern side of the Great Stookan (Fig. 11.7). Proceed east along the path to the Causeway.

The Giant's Causeway is formed of regular near-vertical columns, often hexagonal in cross section and divided by horizontal fractures into ball-and-socket joints (Fig. 11.9). Similar forms can be seen in other localities in the world, notably Fingal's Cave on the Scottish island of Staffa in the Hebrides. However, these regular columns are only part of the picture of jointing that occurs in the Causeway basalts, and the complete jointing pattern is best seen in the cliff exposures 200 m east of the Causeway, at the feature known as the Giant's Organ. Here, the jointing pattern within the flow is seen to consist of the regular near-vertical columns overlain by a narrower and often curved set of columns, which are in turn succeeded by a set of relatively widely spaced, near-vertical columns at the top of the flow (Fig. 11.10).

A Russian geologist working in Ireland, S. I. Tomkeieff, first described this multi-tier jointing pattern in detail. He used terms adopted from classical architecture that are now used to describe such occurrences all

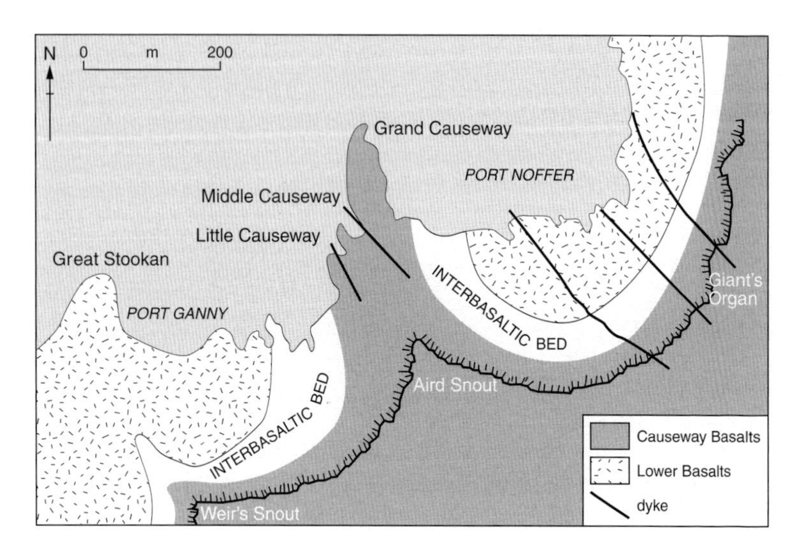

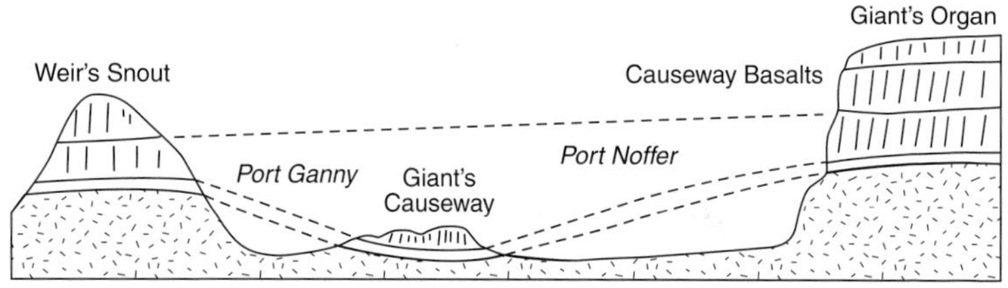

Figure 11.7 Geological map and cross section of the Giant's Causeway locality (after Wilson & Manning 1978).

over the world. The regular columns he called the **colonnade**, the curvi-columnar zone he termed **entablature**, and the upper crudely columnar zone he called pseudo-columnar. Recent workers have referred to a lower and upper colonnade for the regular columns and the cruder upper columnar zone respectively, with the entablature reserved for the irregular curvi-columnar zone between this (Fig. 11.11). The path farther east from the Giant's Organ is at the level of the Interbasaltic laterite and it allows the examination of the weathered Lower Basalt surface. The large rounded patches known as the Giant's Eyes are further examples of spheroidal weathering; in this case the basalt is almost completely altered to laterite (Fig. 11.12). Just around the headland to the east, a narrow dyke can be seen cutting through the basalts from sea level to the top of the cliff. The dyke may well have acted as a feeder or conduit for the lavas at the surface.

Return to the visitor centre by the Shepherd's Path and the cliff-top path, if the weather permits.

Figure 11.8 Basalt sequence at the Giant's Causeway, showing the red Interbasaltic laterite separating the Lower Basalts from the Causeway basalts above.

Figure 11.9 Vertical hexagonal columns with horizontal ball-and-socket joints, Giant's Causeway.

Ballintoy to White Park Bay

From the Causeway, proceed east along the B146 to the A2 and to Ballintoy Harbour (D040455). Car-parking and toilet facilities are available here.

The section of north Antrim coastline from White Park Bay eastwards to Ballintoy Harbour, a distance of about 2 km, contains almost the whole

Figure 11.10 The Giant's Organ, showing the regular columns of the colonnade overlain by the narrower curved columns of the entablature.

range of rock types present along the Causeway Coast, and they are generally well exposed and accessible at all but the highest tide levels. The north coast is defined along here by a major fault, the Ballintoy Fault, part of the Portbraddan Fault (see Fig. 11.1). Many of the geological features in the area show signs of this crustal movement, estimated to be a movement of the fault to the north of about 100 m downwards.

Ballintoy harbour car-park A disused quarry at Ballintoy harbour car-park shows a raised-beach sea cave that has been eroded along the line of weakness represented by the fault. Movement along the fault has clearly broken or brecciated the chalk (see locality 1, Fig. 11.13). The obvious bedding planes seen in the chalk and marked by flint nodules disappear at the fault line, and the rock has a broken and fragmented appearance and is known as fault breccia (Fig. 11.14).

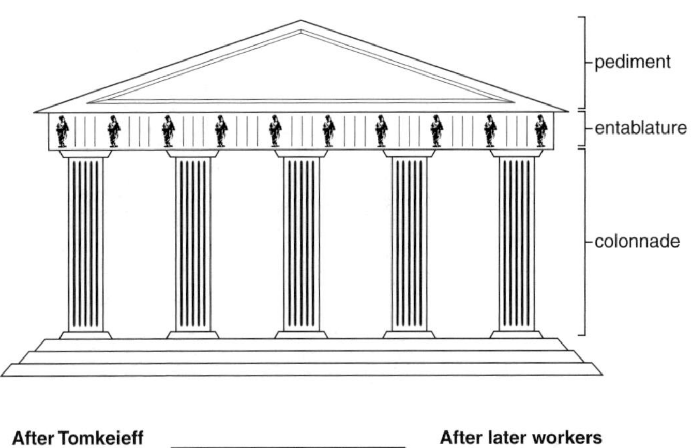

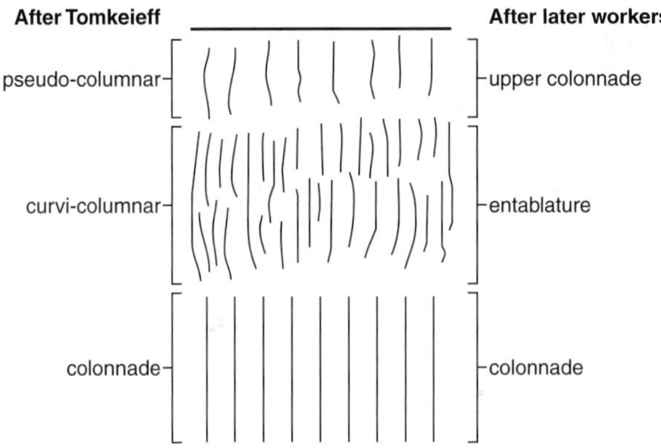

Figure 11.11 The subdivisions of columnar basalts and their origin in classical architecture.

Ballintoy harbour coastal path To the north of the fault zone and just below the high-water mark, in front of the cottages are outcrops of the Antrim Lava Group (see locality 2, Fig. 11.13). A thin bed of pale-green volcanic agglomerate underlies a bright red volcanic dust layer that lies beneath several thin flows of the Lower Basalts. These are in turn overlain by lavas of the Causeway basalts, which show their characteristic jointing pattern on the small offshore islands (Fig. 11.15). Between these two sets of lavas are the Interbasaltic laterites, but they are not exposed above sea level here, and on the geological map their position is inferred. Ballintoy harbour is along the line of the fault, with chalk at sea level on the southern side and basalts at the same level on the northern side (Fig. 11.13).

Ballintoy harbour to White Park Bay

The coastal path to the west reveals clearly the effects of sea-level changes that occurred after the final retreat of the great icecaps 15000 years ago.

Formation of multi-tiered columnar jointing in basalts

The complex jointing pattern shown by the Causeway lavas is predominantly the result of shrinkage as the rock cools. Lavas lose most heat through their top and bottom surfaces, and the internal stresses set up by thermal contraction produce sets of roughly parallel joints or fractures at right angles to the cooling surfaces, along cooling fronts moving down from the surface and up from the base.

Cracks are started at many points on the surface of these fronts, and three-pronged cracks at angles of about 120° occur at each of these points. As the cracks propagate, they intersect to form irregular polygons of between three and seven sides. As cooling proceeds within the flow, the polygonal cracks move into the lava to form three-dimensional polygonal columns. The columns also shrink along their length to form the characteristic ball and socket that divides the colonnade columns horizontally.

The curvi-columnar zone in the Causeway flows is now believed to have been formed because the cooling of the upper part of the flow was modified by the influx of water while it was still cooling. This water passes down into the flow interior via early-formed cracks called master joints and it accelerates the rate of cooling in the upper part of the flow. This rapid cooling forms the entablature, with narrow joints moving down from the surface, while the colonnade of regular columns is being formed by a slow cooling front moving up from the base (e.g. the Giant's Organ).

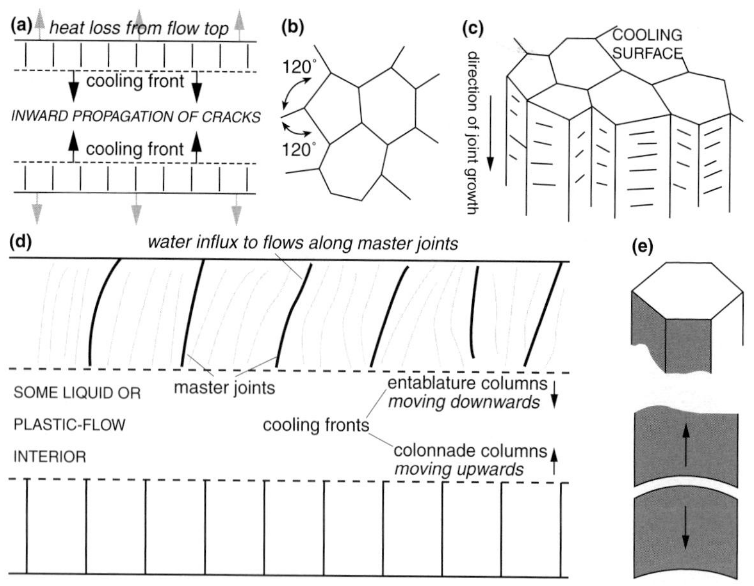

(a) Horizontal cooling surfaces in a lava flow lead to the formation of vertical parallel columnar joints; **(b)** intersecting cooling cracks 120° to each other form irregular polygons; **(c)** polygonal columns form as solidification proceeds and cooling cracks spread downwards; **(d)** multi-tier columnar basalt flows form because of flooding of the flow surface during cooling **(e)** cooling contraction lengthwise along columns forms ball & socket joints that divide the columns vertically.

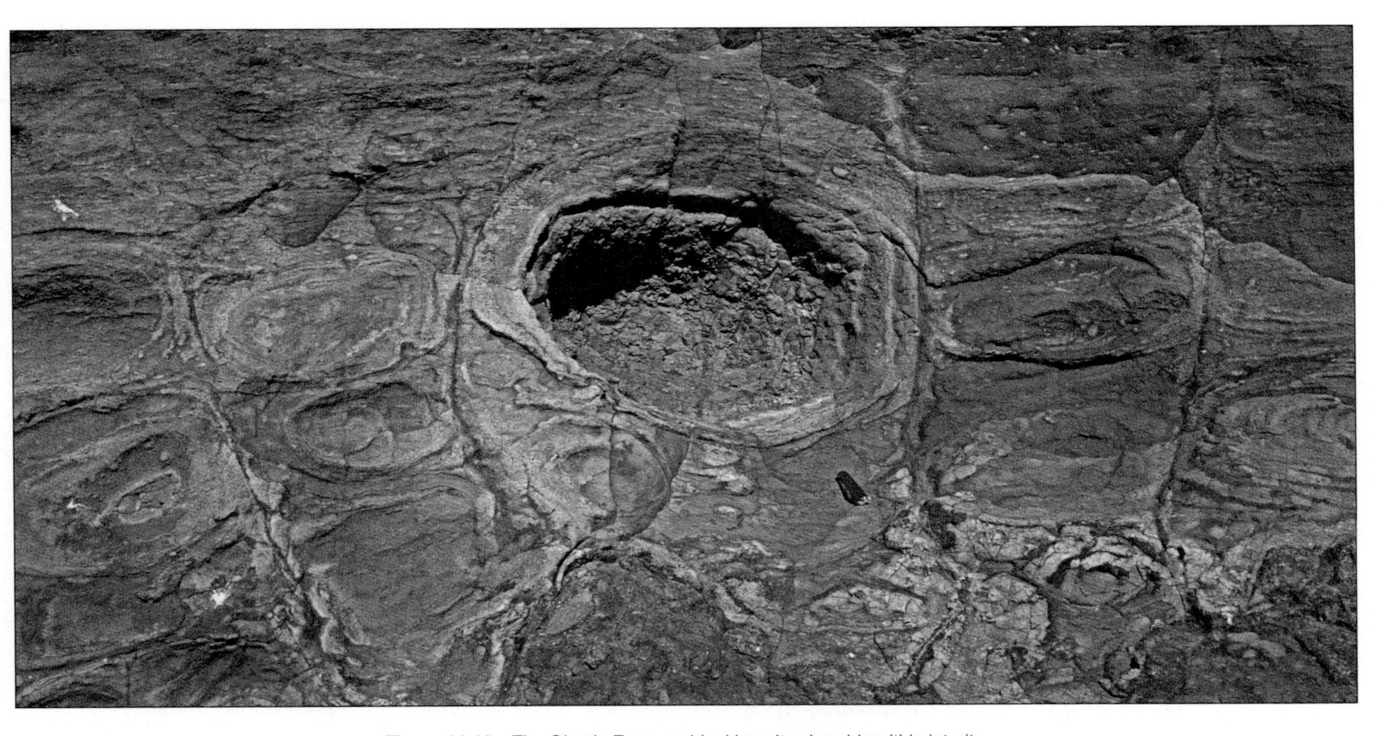

Figure 11.12 The Giant's Eyes: residual basalt spheroids within laterite.

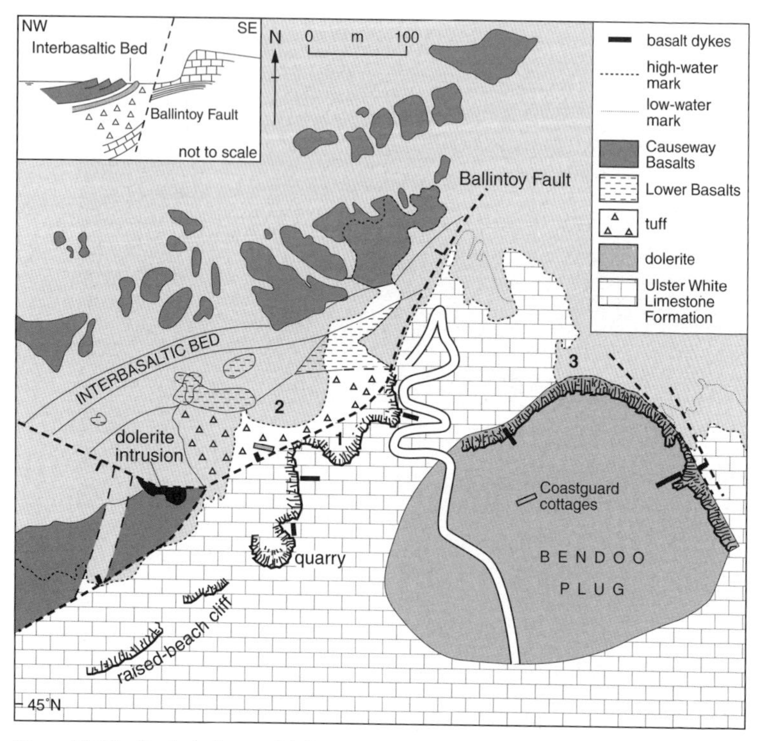

Figure 11.13 Geological map of Ballintoy harbour (after Wilson & Manning 1978). Apart from the small inliers of Lias shales and the igneous intrusions, the area south of the Ballintoy Fault is all underlain by Ulster White Limestone Formation (Chalk).

Figure 11.14 A former sea cave caused by erosion along fault breccia formed along the Ballintoy fault.

Figure 11.15 Lower basalts underlain by red volcanic dust and pale-green volcanic agglomerate.

The enormous thickness of ice during the Ice Age had caused the crust to sink, and recovery from this sinking took longer than the time required for the icecap to melt. The rapid rise in sea level because of the melting of the ice meant that much of this depressed area was temporarily covered by the sea. As the land re-emerged, this former coastline can now be seen as a raised beach some 5–10 m above the present high-water mark. It is marked by a line of former sea cliffs, with several caves, including the one on the fault line exposed in the car-park, and, more obviously, sea stacks and arches, which are now clearly above the present high-water mark.

White Park Bay On the eastern side of White Park Bay the base of the Chalk (Ulster White Limestone Formation) is seen sitting above the Waterloo Mudstone Formation (Lias Clay), which crops out just below the high-water mark, and may be seen if the beach has been scoured clear of sand. As the clay is early Jurassic and the Chalk is late Cretaceous in age, this junction represents an unconformity in the stratigraphical sequence. It also indicates a marked change in sedimentary conditions. The lowest beds of the chalk contain many pebbles and rock fragments, and this represents a basal conglomerate formed as a result of a gradual flooding of the Jurassic land surface by the sea during Cretaceous times. The conglomerates also have a distinctive green speckled appearance because of the mineral **glauconite**, which is indicative of shallow marine conditions. The Lias Clay formed as an offshore marine deposit and is a sticky grey clay with occasional harder dark limestone beds, and contains abundant marine fossils such as ammonites, crinoids and sea urchins. The hummocky ground behind the beach at White Park Bay results from the collapse of the chalk cliffs over the weaker Lias clay.

Bendoo Plug Either walk along the beach to the car-park above the western end of White Park Bay, if you have arranged a pick-up there, or return to the car-park at Ballintoy and examine the Bendoo Plug, just east of Ballintoy Harbour (see locality 3, Figs 11.13, 11.16). This consists of dolerite, a medium-grain basaltic rock, and is roughly cylindrical in shape with a diameter of about 350 m and in steep vertical contact with the surrounding chalk. Contact metamorphism of the chalk can be seen on the beach if there is little sand present. The zone of metamorphism is relatively narrow, about a metre wide. This suggests that the plug was a relatively short-lived intrusion and not a major feeder for the Antrim lavas. Similar features, known as pit craters, form in active volcanic areas such as Hawaii and are temporary lava lakes formed by crustal subsidence. Allow 2–3 hours for the complete excursion from Ballintoy to White Park Bay.

The Carrickarede volcano Just over 1 km east of Ballintoy village is the rope bridge at Carrickarede. Although this precarious connection between the

Figure 11.16 Contact zone between the dolerite of the Bendoo plug and chalk.

mainland and the offshore island of Carrickarede is famous, or possibly infamous, what is less well known is that the island and the area around it are the remains of an extensive explosive volcano of Palaeogene age. The cliffs show clear exposures of agglomerate, consisting of fragments of basalt, chalk and Lias Clay, blown out from a vent that formed in the early stages of the volcanic activity in Antrim. The site is now owned by the National Trust, and car-parking and other facilities are available. The bridge is open from March to September and is used by salmon fishermen. Allow an hour for the visit to the island, longer if the site is busy.

Ballycastle–Fair Head–Murlough Bay

The area of the northeast Antrim coast from Ballycastle, east to Murlough Bay and Torr Head provides some of the most spectacular coastal scenery in Ireland* and a wide variety of geological features, including rocks of Dalradian, Palaeozoic, Mesozoic and Cainozoic ages. The area is dominated by the headland of Fair Head, with its prominent cliffs of columnar dolerite, a promontory featured as early as the second century AD on the classical geographer Ptolemy's map of the world. The excursion aims to examine the various lithologies of the Carboniferous, including the coal seams of the Ballycastle coalfield, the basement schists, psammites and limestones of the Dalradian, and the effects of Palaeogene intrusive activity on the surrounding country rocks (see Fig. 3.1). The excursion is in two parts, starting from Ballycastle and examining first the succession along the foreshore east of Ballycastle and then the rocks around Fair Head, Murlough Bay and Torr Head.

Ballycastle to Carrickmore

Leave Ballycastle eastwards on the A2 road and take the left-hand fork signposted to Corrymeela. Continue east along the coast to the car-park at Marconi's Cottage at D151419, where Marconi supposedly conducted his early experiments in trans-Atlantic wireless communications. This is Colliery Bay, named after the mines used to extract the various coal seams from the Upper Carboniferous sediments that form the cliffs around here. Figure 11.17 shows the detailed geology and stratigraphy of the area. The path to the east passes under cliffs of shales and sandstones, with some thin coals, capped by the 20 m thick Gobb sill, which is Palaeogene (Tertiary) in age. Fallen blocks of dolerite from this intrusion can be examined and sampled on the path. To the west of the car-park on the south side of the road is the entrance to one of the long-abandoned coal mines, the North Star Colliery. This was an example of horizon mining, a method in

* Ballycastle, Sheet 5, 1:50 000 OSNI, Ballycastle, Sheet 8, GSNI.

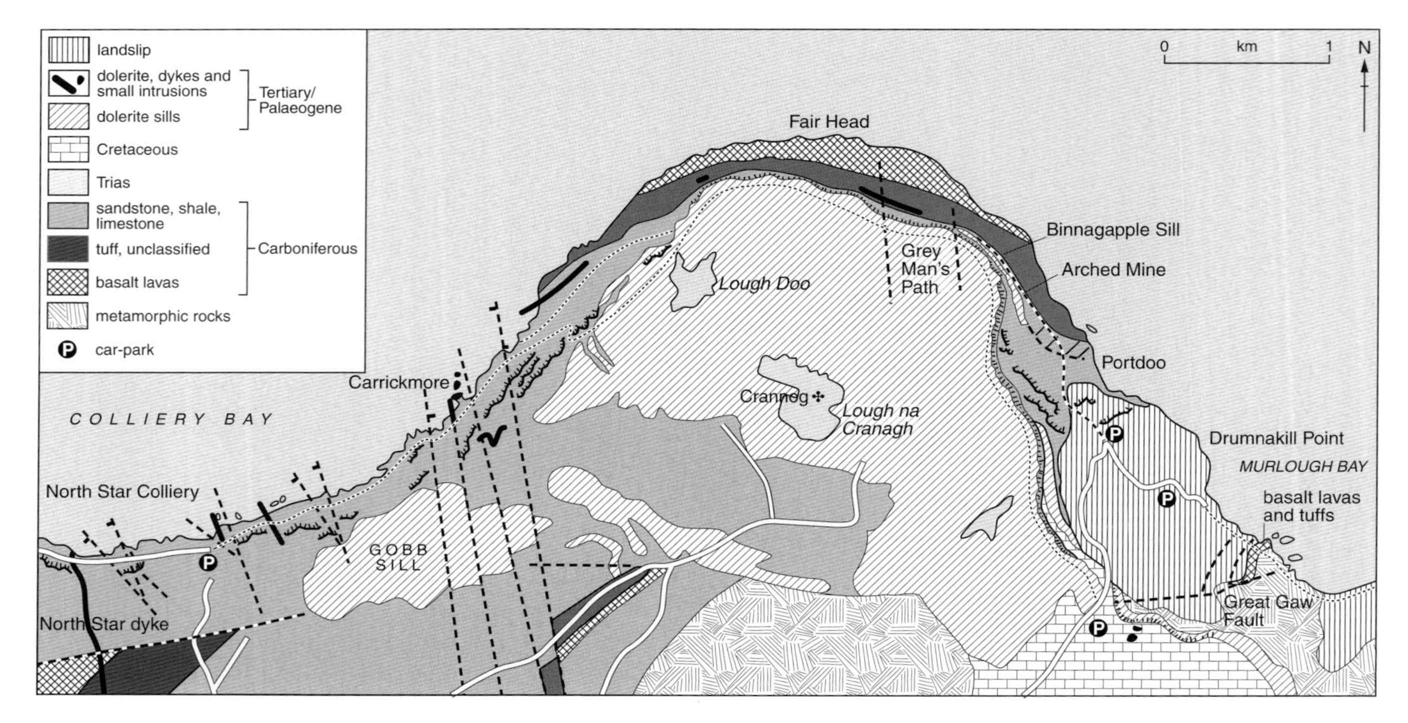

Figure 11.17 Geology of the area around Fair Head (from Emeleus & Preston 1969; after GSNI).

which the coal was worked above the level of the mine and dropped to the haulage wagons through "shoots" or "troughs". The mine was driven with a slight rise inland to the south, allowing the loaded wagons to be pushed out easily and aiding drainage without any need for pumping. Approximately 200 m from the car-park, the North Star dyke is exposed on the foreshore. This is a Palaeogene dyke consisting of olivine-rich dolerite, 4–5 m wide with a northwest–southeast trend. Continue east along the coastal path. At several localities spoil-heaps of black shaly material probably indicate the site of old workings of the Main Coal bed in the Main Limestone. At the head of the cliff the Gobb sill is exposed, intruded into bright red Carboniferous sandstones that dip inland around 15°.

Just before an iron stile crosses a fence, there are outcrops of the Carrickmore Sandstone Group: reddish-brown cross-bedded coarse sandstones and grits containing angular quartz pebbles (Fig. 11.18). A few hundred metres beyond this point lies the small island of Carrickmore, accessible at low tide. Here the seaward continuation of a dyke, exposed on the foreshore above the high-water mark, has intruded Carboniferous shales and fireclays, thermally metamorphosing them to porcellanite (Fig. 11.19). Fireclay is the name given to the soft clay found underlying coal beds and often used in brick making. In addition, the upper part of the intrusion has caused the sediment to dome upwards, forming a blister-like shape. To the east of Carrickmore further exposures of the Carrickmore Sandstone Group are visible, along with the columnar cliffs of the Fair

Figure 11.18 Cross-bedded sandstones of the Carboniferous Carrickmore Sandstone group, Carrickmore.

Head sill and its associated block scree. Access to the western upper part of Fair Head is possible from here by scrambling up the slope behind Carrickmore. Otherwise, retrace steps back to the car-park at Colliery Bay.

From Colliery Bay return to main A2 road and proceed east to Ballyvoy (D156407) and follow signposts for Murlough Bay from the Ballyvoy to Torr Head road. Park in the upper car-park at D191418. The road passes a disused lime kiln en route to the car-park. There are many lime kilns in the area and most of them probably date from the late eighteenth or early nineteenth centuries. The early use of burnt lime was in the manufacture of building mortar, but it became widely used in agriculture as a counter to acidic soils and eventually was also used in the smelting of metal ores. From here there are excellent views over the top of the Fair Head sill across to the Mull of Kintyre and the islands of Rathlin, Jura and Islay. To the east the cliffs are formed of white chalk, the Ulster White Limestone Formation, overlying bright reddish-brown conglomerates, sandstones and marls of Triassic age. These lie unconformably on the Dalradian Schists (see pp. 25–29). Note the spring line at the base of the chalk. There is a display in the car-park explaining the natural history of the area and including way-marked walks on and around Fair Head. The post-Cretaceous age of the sill is evident where it is in contact with chalk in a small quarry just west of the road, below the car-park at D191419. Here a relatively thin lobe of dolerite has recrystallized the chalk. The intrusion of the sill into rocks of Cretaceous age at this locality contrasts with its intrusion into older Carboniferous sediments farther west. This change in the level of the sill is referred to as a transgression.

Figure 11.19 Carrickmore Island, showing thermally metamorphosed shales and fireclays.

Figure 11.20 The Grey Man's Path, Fair Head.

The upper surface of the Fair Head sill Exit the car-park to the west and follow the way-marked trail northwards. This skirts the edge of the cliff, eventually reaching the Grey Man's Path (Fig. 11.20), with spectacular views of the columnar jointing and the associated block scree beneath (Fig. 11.21). Care should be taken close to the edge of the cliff, particularly in poor weather or limited visibility. At the northeast corner of Lough Doo (D173434) there is a 30 m-long exposure of baked or indurated Carboniferous shales. This perhaps represents the baked margin of the sill or even a huge **xenolith** within it.

Either return to the car-park (option 1) or descend the Grey Man's Path to the base of the sill (option 2). If the latter is chosen, extreme care must be taken in the descent, and participants should be properly equipped and suitably experienced hill walkers. Either option will take 5–6 hours.

For option 1, from the upper car-park go down the hill to the next car-park at D192427 and walk westwards along the track. The columnar dolerite of the main Fair Head sill dominates the skyline, but below the

Figure 11.21 Columnar jointing in the Fair Head sill and the associated block scree.

Figure 11.22 Twin entrances to the Arched Mine.

cliffs are extensive exposures of red-brown sandstones and shales of early Carboniferous age, similar to those examined around Colliery Bay on the west side of Fair Head. Occasional thin coal seams here were the focus of many coal mines in the eighteenth and early nineteenth centuries, with their associated spoil heaps, now for the most part obscured by thick vegetation. Around 500 m along this track are the twin entrances to the Arched Mine (Fig. 11.22). This was one of the larger mines in the area, as is reflected in the relatively sophisticated nature of the mine opening. The coal extracted here was of generally poor quality, but was used to raise steam in stationary steam engines and also to fire the lime kiln noted earlier on the road near the top car-park and another kiln on the shore at Murlough Bay. The ruined cottages to the south of the mine entrance date from this early period of coal mining here. Below the Fair Head sill here a second subparallel sill, the Binnagapple sill, can be recognized (Fig. 11.23). The columnar jointing that formed as the sills cooled makes them susceptible to weathering by ice expansion as water percolating downwards expands on freezing. This dislodges large blocks from the cliff face, producing the remarkable tangle of huge rectangular dolerite blocks that forms the scree below the sill all around the headland (Fig. 11.24). The path westwards along the base of the cliff and above the block scree shows the base of the Fair Head sill in contact with the underlying Carboniferous sandstones and shales. The block scree is difficult and dangerous to cross. The path around the base of the cliff leads eventually to the lower end of the Grey Man's Path. Retrace your route to the car-park.

Figure 11.23 Fair Head sill with underlying subparallel Binnagapple sill.

Figure 11.24 Oblique aerial view of Fair Head sill, showing the gently inclined upper surface with Lough na Cranagh on the left and the extensive block scree that mantles the cliffs, formed by frost action on the near-vertical columnar jointing (courtesy of Nigel McDowell).

For option 2, proceed down the Grey Man's Path to the base of the sill. Walking eastwards along the track above the block scree, the base of the sill, in contact with the underlying sandstones and shales, can be examined. Follow the track to the Arched Mine and eastwards to the car-park at D192427, the starting point for option 1.

Murlough Bay foreshore Proceed east along the road to the lower car-park at D196423. There is no access for vehicles beyond this point. The promontory of Drumnakill Point, about 200 m north of the car-park, is formed by a mass of columnar dolerite that has slipped from the upper levels of the bay. Little appears to be known about the ruins of a small church on the headland. A short distance along the track on its seaward side is another lime kiln, similar to the one at the top of the bay. This lower kiln is made of sandstone blocks, unlike the one above, which is basalt or dolerite. On the foreshore in front of the first cottage (the Bothy) to the east are remnants of lavas and ashes of Carboniferous age, much older than the Palaeogene lavas exposed elsewhere in Antrim, or the sills of Fair Head. The foreshore in front of the second cottage consists of Carboniferous sandstones and grits, but the geology changes abruptly eastwards across the bay to convoluted folded schists of Dalradian age. This contact is faulted (see Fig. 11.17): the Great Gaw Fault runs approximately east–west from Murlough Bay to Ballycastle Bay, where it terminates against the Tow Valley Fault. It is thought that the rocks to the north of the fault have moved down by as much as 400 m in places.

Further examples of Dalradian rocks are exposed at Torr Head, about 5 km around the coast to the southeast from Murlough Bay.

Return to the car-park.

Torr Head From the Murlough Bay car-park, return to the Ballyvoy–Torr Head road at D186405 and proceed east to Torr Head, parking in the car-park at D233404. The final part of this road is very steep and narrow, and is not suitable for large vehicles.

The quartz schists of the Murlough Bay Formation are the oldest Dalradian beds exposed in northeast Antrim, part of the Argyll Group of the Dalradian Supergroup. At Torr Head are two younger formations, the Torr Head Limestone Formation and the overlying Altmore Formation, which consists mostly of coarse-grain psammites and grits. At Torr Head this succession is upside down because of extensive folding; the overturned contact between the two formations can be seen on the south side of the headland. From the top of the sea cliff southeast of the abandoned coastguard station, a descent *down* the slope actually passes *up* the succession, from the limestone to the younger psammites (Fig. 11.25). Here the

Figure 11.25 Torr Head Limestone Formation in the background, structurally above but stratigraphically below the psammites of the Altmore Formation in the foreground.

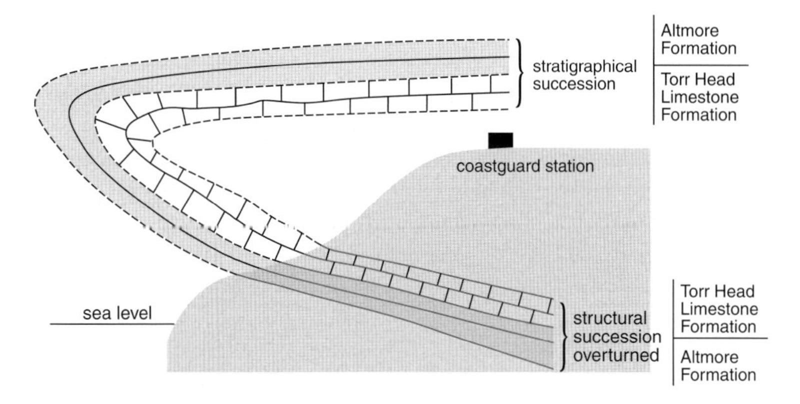

Figure 11.26 Hypothetical cross section through the east side of Torr Head, showing one possible explanation for the overturned succession in the area.

limestone is up slope from the psammites in the foreground, it is therefore *structurally* above the psammite beds, but is *stratigraphically* below them. Figure 11.26 shows one possible explanation for the overturned succession at Torr Head, with the two limbs of a large and very tight fold almost parallel to each other. The lower limb is inverted, producing the rock sequence seen at Torr Head. The main headland is composed of an intrusive body of metamorphosed basic igneous rock of Dalradian age that probably represents a sill intruded into the sediments before metamorphism.

Return to the car-park. Allow an hour for this part of the excursion.

*Mid-County Antrim**

Introduction

South of the Tow Valley Fault and the Precambrian metamorphic rocks of the Murlough Bay–Torr Head area, the geology of mid-Antrim is marked by mainly Mesozoic rocks of Triassic and Jurassic age, which underlie the Cretaceous chalk and the thick succession of Palaeogene basalts that comprise the Antrim Plateau. Around Cushendun and Cushendall are some important outcrops of Devonian rocks that represent the debris from erosion of the Caledonian mountains in a hot desert environment some 400 million years ago. South from these desert sandstones and conglomerates, the landscape is dominated along the coast by the spectacular cliffs of the Antrim coast road. Here the dark basalts contrast starkly with the white

* OS Sheet 15 Belfast, Sheet 9 Larne, Sheet 5 Ballycastle; GSNI Sheet 21 Larne (solid and drift), Sheet 29 Carrickfergus, Sheet 8 Ballycastle

chalk cliffs of the Ulster White Limestone Formation beneath, which in turn is underlain in many places by the grey clays of the Jurassic Waterloo Mudstone Formation, also known as the Lias Clay (Fig. 11.27). The Waterloo Mudstone is the source of many and varied fossil species but also the cause of the inherent instability shown by the coast road, which has been prone to major landslips in many places. The effect of the geology on this major transport link in northeast Ireland will be examined in the course of the excursion. It also aims to examine the relationship between the surface of the limestone and the overlying basalts, and consider the change in geological environment represented by such a junction. The detailed characteristics of the Lower Basalts are clearly shown along the coastal footpath from the town of Whitehead towards Black Head some 2 km to the north.

Whitehead to Larne

Proceed to Whitehead and park in the car-park on the promenade at J479922. Take the path along the coast northwards towards the lighthouse at Black Head (J488934) to the start of the outcrops of Lower Basalts. The basalt flows beside the path show typical flow structures, including reddened flow tops with zeolite-filled gas bubbles, and pipe vesicles at the base of flows and flow units (Fig. 11.28). In one of the thicker flows near where the path climbs up to the lighthouse, a more massive flow shows well developed vesicle cylinders (Fig. 11.29). These are parallel lines of gas bubbles or vesicles that rose through the lava flow as it stagnated and cooled.

At the first bridge on the path is a sea cave formed by erosion along a 5 m-wide dolerite dyke running northwest–southeast along the typical Palaeogene or Tertiary direction. At the second bridge is a good example of a red flow top of a blocky lava surface. On the foreshore north of the end of the path is an outcrop of coarse agglomerate material consisting of large blocks of basaltic material in a finer reddish matrix (Fig. 11.30). This probably represents a small feeder vent and, like the dyke noted earlier, has a northwest–southeast trend. Return to the car-park either by retracing the route along the coastal path or via the lighthouse at the top of the cliff.

Waterloo Foreshore, Larne From the promenade at Whitehead take the road signposted to Islandmagee and Larne, and follow the A2 to Larne. In Larne follow signs for the harbour and turn off the A2 at Curran Road. Turn left off Curran Road to the Leisure Centre and proceed to the car-park on the sea front. The foreshore here shows the contact between Upper Triassic rocks, the Mercia Mudstone and Penarth groups, with the overlying Lower Jurassic, the Waterloo Mudstone Formation. The Mercia Mudstone Group consists of red siltstones becoming grey-green mudstones up the

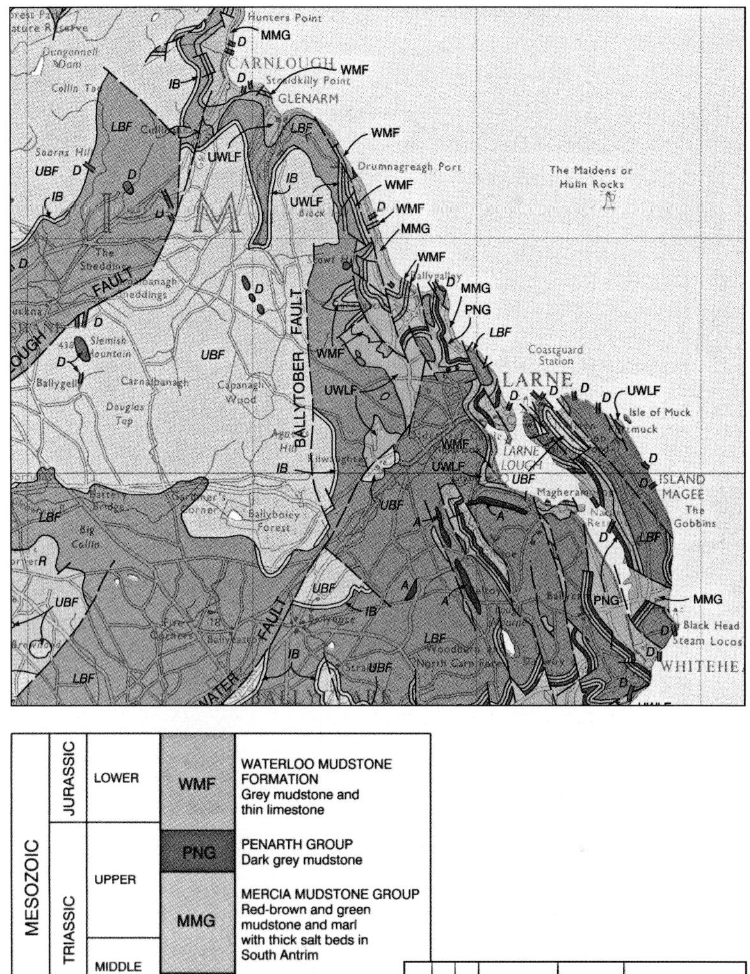

MESOZOIC	JURASSIC	LOWER	WMF	WATERLOO MUDSTONE FORMATION Grey mudstone and thin limestone
MESOZOIC	TRIASSIC	UPPER	PNG	PENARTH GROUP Dark grey mudstone
MESOZOIC	TRIASSIC	UPPER / MIDDLE	MMG	MERCIA MUDSTONE GROUP Red-brown and green mudstone and marl with thick salt beds in South Antrim
MESOZOIC	TRIASSIC	LOWER	SSG	SHERWOOD SANDSTONE GROUP Red-brown sandstone
UPPER PALAEOZOIC	PERMIAN	UPPER	BELF	BELFAST GROUP Marl with gypsum, dolomitic limestone, basal sandstone
UPPER PALAEOZOIC	PERMIAN	LOWER	ENLE	ENLER GROUP Red-brown sandstone, conglomerate and siltstone, basal breccia

UPPER PALAEOZOIC	DEVONIAN	UPPER	FAMENNIAN		
UPPER PALAEOZOIC	DEVONIAN	UPPER	FRASNIAN	RAF	RED ARCH FORMATION Red conglomerate and sandstone
UPPER PALAEOZOIC	DEVONIAN	MIDDLE	GIVETIAN		
UPPER PALAEOZOIC	DEVONIAN	MIDDLE	EIFELIAN		
UPPER PALAEOZOIC	DEVONIAN	LOWER	EMSIAN	CROS	Cushendall Porphyry CROSS SLIEVE GROUP Purple conglomerate and sandstone
UPPER PALAEOZOIC	DEVONIAN	LOWER	SIEGENIAN	CROS	
UPPER PALAEOZOIC	DEVONIAN	LOWER	GEDINNIAN		

Figure 11.27 Geology of the mid-Antrim area (© Crown copyright).

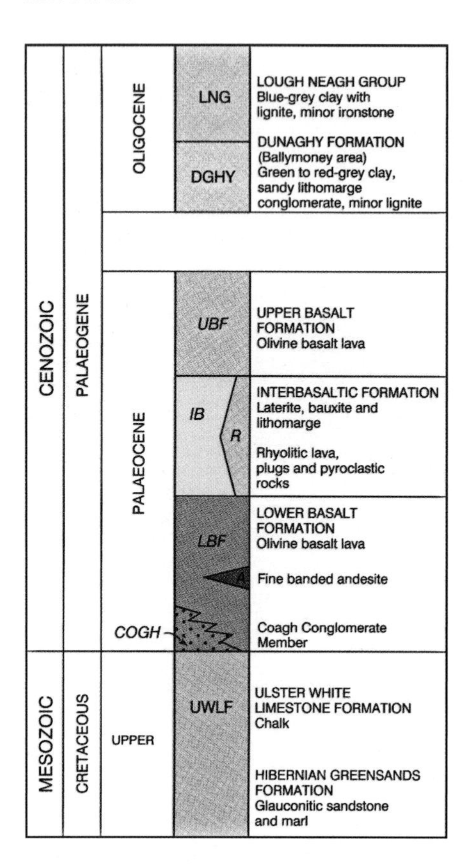

Figure 11.27 continued Third part of the keys to the map on p. 181 (© Crown copyright).

succession. The overlying Penarth Group and the Waterloo Mudstone Formation are grey-black mudstones, and the Triassic/Jurassic boundary is taken as the first appearance in the succession of the ammonite *Psiloceras*. This occurs near the sewage outfall pipe seen on the foreshore. The change in colour of the sediments reflects the gradual transition from a predominantly terrestrial environment in the Triassic to a marine environment in the Jurassic.

Mud flow at Drumnagreagh From the foreshore at Waterloo retrace the route back to the A2 and proceed north along the coast following the signs to Glenarm. Proceed to a small car-park on the seaward side of the road north of Drumnagreagh Port at J341134. The car-park is approximately 500 m north of the sign for the Drumnagreagh Hotel. The mudflow in the Waterloo Mudstone Formation (Lias Clay) can be seen on the landward side of the road 200 m north of the car-park (Fig. 11.31). The Waterloo Mudstone occurs beneath substantial thicknesses of chalk and basalt,

Figure 11.28 Flow structures, including reddened flow tops, zeolite-filled gas bubbles and pipe vesicles; coastal path, Whitehead, County Antrim.

which form the spectacular cliffs elsewhere on the coast road. However, the mudstone forms an unstable foundation for these cliffs, especially during periods of heavy rainfall, with the result that it squeezes out from beneath the cliffs like toothpaste from a tube, and major landslips then occur. This means that keeping the road open and safe has required some

Figure 11.29 Vesicle cylinders, Black Head, County Antrim.

Figure 11.30 Vent agglomerate, Black Head, County Antrim.

major civil-engineering projects and the road still requires constant monitoring and repair. The A2 can be extremely busy at times, so care should be taken when approaching the site of the mudflow.

Chalk-basalt contact at Whitebay near Glenarm Proceed north on the A2 approximately 2 km to the car-park and picnic area at Whitebay (J328152). A prominent disused quarry on the landward side of the road clearly shows the contact between the chalk and overlying basalt (Fig. 11.32). The chalk surface was clearly deeply weathered by karstic solution before

Figure 11.31 Mudflow in Waterloo Mudstone (Lias Clay), Drumnagreagh, Antrim coast road.

Figure 11.32 Basalt/chalk contact at Whitebay, near Glenarm, County Antrim.

eruption of the basalts and there are deep solution hollows on the chalk surface, which were infilled by the advancing lava flows. A bright red layer clearly visible between the whitish chalk and the grey-brown basalt is usually described as residual "clay with flints", left when the chalk is dissolved away by rain water and groundwater. However, the Antrim chalk has an almost negligible clay content, and the reddish material here is more likely to be weathered volcanic ash that buried the Cretaceous land surface in the early stages of the succeeding Palaeogene volcanism. The quarry also shows the near-horizontal bedding planes in the chalk, marked by lines of flint nodules. There are still many loose blocks on the quarry floor, where these features can be examined and there is no need to approach the quarry face.

Reconstruction of the Antrim Coast Road From Whitebay proceed approximately 500 m north towards Glenarm to a picnic site on the seaward side of the road at J323155. Park here and walk north towards Glenarm. In 1967 a major rockfall occurred on this section of the road, completely blocking it and necessitating the closure of the road. This section of the road had been built in the 1840s to provide employment for the local population during the years of the famine. It was proposed to construct a new causeway, seawards of the original undercliff road and protected by a bank of rock armour composed of limestone blocks, many of which come from the disused quarry at Whitebay (pp. 184–185). Figures 11.33 and 11.34 show the line of the road as it currently appears and a cross section through the

Figure 11.33 The reconstructed Antrim coast road near Glenarm, showing the new causeway and the protective rock armour.

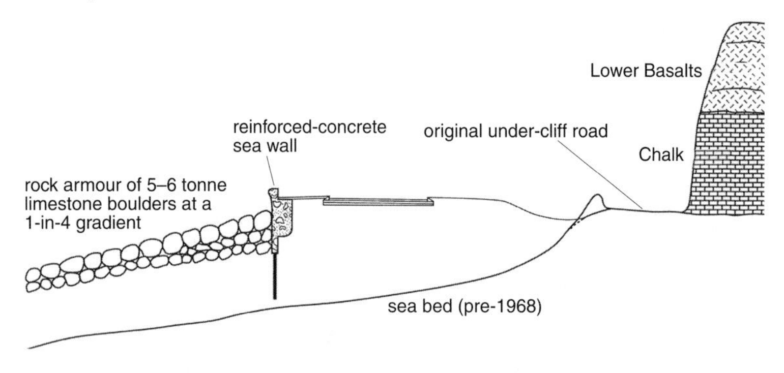

Figure 11.34 A cross section through the reconstructed causeway.

reconstructed causeway. The low concrete wall visible in the centre of the photograph is reinforced and supported on piles to a depth of 4m. The rock armour totalled around 100000 tonnes, with blocks weighing 6–15 tonnes, and each block being placed individually by crane.

The Antrim coast road provides a good example of the conflict that often occurs between the demands of the communication network, in this case the major east-coast route in Antrim, and the problems set by geological factors, in this case an unstable layer occurring beneath massive cliffs of basalt and chalk at the precise level of the road.

Rotational shears at Garron Point The final example of the effect of the unstable clay layer underlying the basalt and chalk cliffs on the Antrim coast road is to be seen at Garron Point. Proceed north through Glenarm and Carnlough villages to the car-park and picnic site on the seaward side of the road just south of Garron Point at J302238. Looking north to Garron Point (Fig. 11.35) the slumped blocks of basalt and chalk are clearly visible, showing the characteristic steep inland dips associated with this type of landslip. As the underlying clay layer fails, the overlying chalk and basalt slump along a curved plane, and the phenomenon is referred to as rotational slip (Fig. 11.36). It is thought likely that this phase of slumping occurred at the end of the Ice Age when the ice sheets melted relatively quickly and the cliffs were left without support. The dwellings by the roadside, north of the car-park, are former coastguard cottages that were associated with a lookout point on Garron Point above.

Redbed sedimentary rocks of northeast Antrim The coastal strip between the villages of Cushendall and Cushendun has excellent exposures of sedimentary rocks (sandstones and conglomerates mainly), which are often referred to as redbeds. These are predominantly red because of an iron

Figure 11.35 Slumped blocks of basalt and chalk at Garron Point, caused by rotational slip.

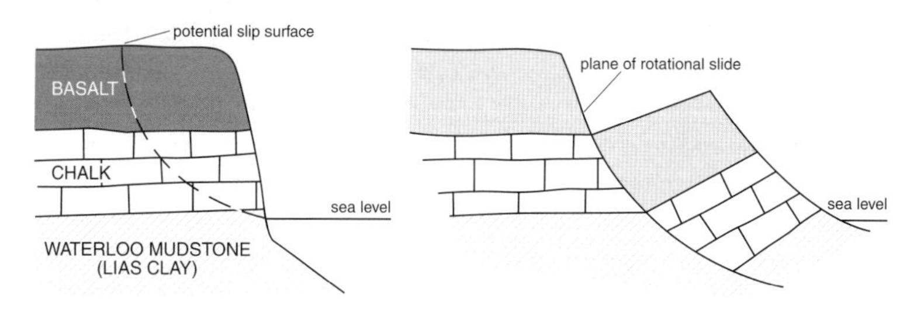

Figure 11.36 Cross sections through Garron Point, showing the process of rotational slip.

Table 11.1 The stratigraphy of the Devonian rocks of northeast Antrim.

Upper Devonian	Red Arch Formation Red conglomerates and sandstones
Middle Devonian	Absent
Lower Devonian	Cross Slieve Group Purple conglomerate and sandstone

oxide, haematite, which covers individual grains and fragments of the rock. These sediments were deposited in a continental desert environment far from any marine influence. On the 1:250000 map (see Fig. 6.6) all of these beds are considered to be Devonian in age (Table 11.1), but there is at least the possibility that some of the rocks around Red Bay, south of Cushendall, are much younger, possibly Triassic in age. Without either conclusive palaeontological or radiometric evidence, it is one of those geological enigmas that awaits a definitive solution. The excursion will examine the oldest sediments first, beginning at the northern end of the outcrop at Cushendun, which are considered to be of undoubted lower Devonian age.

From Garron Point continue north on the A2, through Cushendall towards Cushendun. Follow the signs off the A2 on to the B92 for Cushendun. Proceed to the car-park at the southern end of the village near the Bay Hotel. Follow the path north and east of the Bay Hotel on the coast, where thick beds of boulder conglomerates are exposed in the cliffs and sea caves that lead to Cave House (D253326). The sediments dip southeast and rest unconformably on Dalradian metamorphic rocks. These beds of the Cushendun Formation contain abundant quartzite pebbles and boulders with occasional thin sandstone or mudstone beds (Fig. 11.37). The sediments were produced in alluvial fans in a desert environment, with the material derived from the northwest. Occasional occurrences of mudstone with desiccation or drying cracks confirm the hot arid climate, with periodic rainfall followed by periods of drying out.

Port Obe and exposures of Lower Devonian, the Cross Slieve Group Retrace the route to the Bay Hotel, proceed south to Knocknacarry (D241322) and take the road to the south to Port Obe (D247288). Park in the layby for Layd church, a former parish church for the area, which was closed in the eighteenth century and has been used as a burial site since then. Take the coastal path to the south and turn off onto the narrow track down to the foreshore at Port Obe. The rocks exposed are purple-red coarse to medium-grain sandstones with some cross bedding visible in places. Two green mudstone beds exhibit desiccation cracks infilled with sandstone from the succeeding bed. Retrace the route to Layd church and proceed to Cushendall.

Figure 11.37 Conglomerates of the Cushendun Formation (Devonian), near the Bay Hotel, Cushendun, County Antrim.

Cushendall–Waterfoot shore section

Proceed to the car-park at the lifeboat house (D244269) on the east side of the road just south of Cushendall. The localities are best visited at middle to low tide levels; the section involves walking south along the foreshore to the Red Arch (D243261) and returning to the car-park via the footpath on the main road (A2). The road trip is approximately 1.5 km. Allow 1–2 hours for this section of the excursion.

The aim of this section of the excursion is to examine the sediments of the Red Arch Formation, designated here as Upper Devonian, which are in unconformable contact above the Lower Devonian rocks of the Cross Slieve Group examined at Cushendun and Port Obe (see Table 11.1). In the absence of palaeontological evidence, it has been suggested that they may be younger than Devonian, and perhaps of Triassic age, on the basis of similarities to Triassic rocks in the west of Scotland.

The wavecut platform at the lifeboat house shows conglomerate beds and coarse sandstones of the lowest part of the Red Arch Formation, with some boulders up to 1 m in diameter and occasional beds of coarse sandstone with pebbles. An unconformity separating the conglomerates of the lowest part of the Red Arch Formation from the upper parts is conspicuous on a small island about 200 m south of the lifeboat house at D244267, opposite the caravan park on the west side of the road (Fig. 11.38). It is marked by a change upwards from fine to medium-grain conglomerates and pebbly sandstones to very coarse clast-supported conglomerates that

Figure 11.38 Unconformity between the lower and upper parts of the Red Arch Formation, near the Lifeboat House, Cushendall, County Antrim.

dip at a lower angle than the underlying rocks. Depending on interpretation, this unconformity is either between Devonian rocks of differing ages (intra-Devonian) or is the junction between the Devonian and Triassic periods.

Coastal exposures 200 m north of the Red Arch above the conglomerates examined in the previous locality are pinkish-red pebbly sandstones with some cross stratification. The environment of deposition was probably a shallow stream with gravel bars (Fig. 11.39).

The Red Arch Formation sediments probably represent deposition on

Figure 11.39 Pinkish-red pebbly sandstone of the Red Arch Formation, near the Red Arch, Cushendall, County Antrim.

an alluvial fan, with the material being re-worked at a later stage by braided streams to form the pebbly, often cross-bedded, sandstones. Return to the car-park via the footpath along the A2 coast road.

Slemish and the laterite mines of mid-Antrim Given time, opportunity and a suitable means of transport there are some interesting geological features that may be examined and explored within about 25 km from Cushendall. Take the B14 road southwest from Cushendall; this road climbs through Glenballyemon, between the spectacular cliffs of Lurigethan, with its summit promontory fort to the south, and the peak of Tievebulliagh, site of the famous axe factory, to the north (see p. 72). Turn off the B14 at D186205 onto East Skerry Road. Approximately 2 km along this road near Legagrane are the remains of some of the mines and spoil heaps of the iron-ore industry that was based on the Interbasaltic laterite beds of the Antrim Lava Group. The hill to the north is Slievenanee, which means "iron mountain", suggesting an early appreciation of the use to which the laterite could be put. In places in the area are remains of the narrow-gauge railways and tramways that were used to transport the ore in the late nineteenth century; although most was smelted in Britain, the village of Newtown-Crommelin (D143175) still has the remains of a peat-fired smelter set up in the middle of the eighteenth century. A definitive account of the mid-Antrim mining industry is to be found in Hamond (1991).

From Newtown-Crommelin, follow the signs to the A43 to Ballymena and Broughshane, and proceed to Broughshane. Take the B94 to Ballyclare and take the first road to the left, signposted to Slemish (brown sign). Follow the sign to Slemish, which is a volcanic plug of Palaeogene age with Hawaiian lava-lake affinities. The hill forms a prominent topographical feature in mid-Antrim and is best known for its association with Saint Patrick and the early Christian Church in Ireland. The visitor centre has information on the geology, archaeology and local history of the area. There is a way-marked path to the summit, a climb of about 180 m, which takes about an hour for the round trip.

Appendix

The Earth Science Conservation Review

Conserving the Earth heritage resources of Northern Ireland

I. J. Enlander

Many of the sites described in this book, which lie within the six counties of Northern Ireland, have one thing in common: they have all been identified as being among the most important geological localities in that geographical area. A research programme, the Earth Science Conservation Review, is being undertaken by the Department of the Environment within Northern Ireland. The purpose behind this exercise is a systematic evaluation of sites of geological and geomorphological significance, ultimately to identify those sites that collectively best tell the history of this part of planet Earth.

What is the Earth Science Conservation Review? Why is it being undertaken?

Environment and Heritage Service (EHS) is an agency within the Department of the Environment. It is the body charged with giving advice to government on environmental and conservation matters in Northern Ireland. Part of its statutory duty is to identify and protect areas that are of special scientific interest because of their biological or Earth heritage features.

The Earth Science Conservation Review (ESCR) is a programme of research undertaken by EHS to identify Earth science sites of importance in Northern Ireland. The wide range of sites covers both hard rock and process sites, which are between Precambrian and Recent in age. Fifty-two themes have been identified and reported on, producing a list of some 350 discrete features of significance. Some sites host more than one significant feature, reducing the number of actual sites to some 300.

Although there had been some earlier progress in identifying and protecting a range of Earth heritage localities, this had not been done systematically and it was lacking a strong element of site comparison. Such a methodical and standardized approach is needed, given the increasing frequency of challenges to the site selection and designation processes in both the UK and Ireland. Ultimately, the selection of a particular site may have to be defended in a court of law or at a planning inquiry. Selection justifications must therefore be robust.

Earlier legislation (Amenity Lands Act (Northern Ireland) 1965) provided the means to designate Areas of Scientific Interest (ASI); 47 of these were confirmed, of which 26 were selected primarily or partly for their geological or geomorphological interest. While ASIs are still current, no additional ones are being declared, as the empowering legislation was superseded in 1985. The ASI status of these sites ensures that the relevant authority is consulted if a planning application is received for any proposed development that could affect them. The principal weakness of this designation relates to its inability to influence activities, especially those of landowners, for which planning permission is not a requirement, but which could impact on the site's scientific interest. In addition, there was no systematic approach to site selection. However, some of these ASIs have been re-designated under more recent legislation, confirming their scientific merit.

This more recent legislation, the Nature Conservation and Amenity Lands (Northern Ireland) (1985) Order, as amended in 1989, places on EHS the duty to identify and protect the most important geological and geomorphological sites in Northern Ireland. It requires that

"Where the Department . . . is satisfied that an area of land is of special scientific interest by reason of its flora, fauna or geological, physiographical or other features . . . the Department shall make a declaration that the area is an Area of Special Scientific Interest."

This statutory requirement emerges from the recognition that such Earth science sites are important for several reasons. There is an international obligation (not legislatively based) to protect and retain sites that have played a significant role in the development of the science or which have an importance beyond our geopolitical border (Malvern International Task Force 1995). A wider need relates to the provision of localities for present and future research, teaching and training. The network of protected sites also provides the basis of an historical story that has value in predicting future environmental change. This is particularly true of Recent localities that contain extended climatic records in peat bogs, and the more episodic records in dunes and caves. Those sites are often among the most vulnerable to human impacts and they highlight the need for methodical and systematic protection. Many Earth heritage localities are also a significant recreational, landscape and ecological resource.

There is also a broader recognition of the need for Earth heritage conservation as a moral process, in parallel with the belief among some biological conservationists that species and habitats quite simply have a right to exist (Rights of the Memory of the Earth 1991).

The need to protect also emerges from the very real threats that Earth science sites may be under, which can obliterate features of importance, interfere with the natural processes required for their continued existence, or detract from the integrity of landscape. In Northern Ireland, a region where inland outcrop is often very limited because of blanketing by glacial drift, quarries, both active and disused, are a very important source of information on hard-rock geology. However, Northern Ireland also has the highest proportion of domestic and industrial waste returned to landfill in the UK and the continuing demand for landfill sites represents a continuing threat to these quarries.

Additional threats come from coastal developments and sea defences, impacting on both natural coastal exposures and coastal process sites. Landform complexes, particularly deglacial sand and gravel systems, are vulnerable to aggregate extraction, and large-scale landscape changes, as with blanket afforestation, can obscure features and limit access. Finally, the geological community, both amateur and professional, can also have undesirable impacts on vulnerable sites, notably from overcollecting. The degree of impact from collectors is related to site extent, accessibility and frequency of use. The ESCR presents a useful opportunity to evaluate site vulnerability and, if necessary, redirect casual or teaching groups to more robust localities in order to retain critical sites for demonstration and research purposes only. These sites are of finite capacity and must be used in a manner that maximizes their useful life (Nature Conservancy Council 1990).

The structure of the ESCR

It was recognized that the process for site selection must be rigorous, in order to produce a site list that would be meaningful and also defensible against any legal challenge. The tried and tested structure and process of the Geological Conservation Review (Ellis et al. 1996), as undertaken in Britain, appeared to be a viable format to adopt for Northern Ireland.

The Earth Science Conservation Review (ESCR) process can be summarized as follows:

- establishing the subject block series within which individual sites are assessed
- defining the site selection criteria
- defining operational criteria
- appoint individuals or groups to select, assess and report on sites
- review theme literature and shortlist sites
- field visits and individual site investigation
- final assessment and site reports, including summary and fuller descriptions, together with site map and boundary defining the area of interest.

Although in the main the ESCR followed the model set by the GCR, there are some notable variations. Naturally, the geological framework within which sites were to be assessed differed, reflecting the geological character of Northern Ireland, for instance with its truncated Jurassic and Cretaceous stratigraphies. The selection and operational criteria remained broadly unaltered (see below). However, a further difference lies in the reporting procedure. Given the fact that Earth heritage conservation has lagged behind the programme in the rest of the UK, the decision was made to undertake site selection and full site reporting simultaneously, whereas these processes were often separated by many years in Great Britain. In Northern Ireland, the site selection and documentation process is now completed.

Given the particular role of EHS as statutory environmental consultee within the government of Northern Ireland, some aspects of the ESCR were extended to allow EHS to fulfil this wider advisory role more fully. Thus, although site selection focused on nationally and internationally important sites, a first attempt was also made to identify locally significant geological sites. Such sites may not be designated as Areas of Special Scientific Interest (ASSI), but may well serve as the basis for a Regionally Important Geological and Geomorphological Sites (RIGS) network, currently absent from Northern Ireland.

In several instances EHS felt that the knowledge base available for assessing certain ESCR themes was very limited. Where this related to site types deemed to be highly vulnerable, site selection was preceded by a full evaluation of the national (Northern Ireland) resource, as was the case for the karst geomorphology and the deglacial sand and gravel complexes of Northern Ireland. This then allowed an informed selection of key sites, while also providing a comprehensive dataset with wider applications, notably for use in strategic planning and development consultations.

The key objective of the ESCR is systematically to define and protect all Earth science localities in Northern Ireland that are of at least national (the land unit of Northern Ireland) significance.

The general process is as follows.

A thematic framework was established, within which sites were assessed (see Table 1).

Site selection has been undertaken by acknowledged experts within the particular field. Criteria were set down to aid with the selection process. Examples of criteria are:

- sites must contain features that are of at least national importance.

Preference to be given to sites (selection criteria):

- with an assemblage of features or interests
- showing an extended or near complete record
- with intact geomorphological features
- with a sound research history and with good potential for future study
- that have played an important role in the development of Earth science

Table 1 Thematic structure for the ESCR.

Theme	Blocks
Stratigraphy	24
Igneous petrology	5
Palaeontology	8
Structural/metamorphic	5
Mineralogy and metallogenesis	1
Pleistocene	4
Recent	5
TOTAL	52

In addition, the selected suite of sites (operational criteria) should show a minimum of duplication in the hosted features – this is a fundamental difference between the Earth science and biological conservation programmes must be practicably conservable.

Where applicable, a network approach was taken whereby individual sites were viewed in the context of their relationship with other sites in that network, the value of each site being enhanced by this mutuality. Examples of such networks include palaeogeographical sedimentary basins (e.g. the Carboniferous Armagh Basin) and also igneous complexes (e.g. the Slieve Gullion Tertiary Complex). This approach ensured that the network was being described using the minimum number of necessary "subsites", to represent the sites diversity.

Selected sites were then documented and mapped to a standard format. Following a site review, some sites will be designated as Areas of Special Scientific Interest to produce a site network essential to the demonstration of the geological and geomorphological development of Northern Ireland. Table 2 summarizes the number of features selected and the progress made to date in designating the selected sites. The term "feature" refers to the geological phenomenon for which the site has been selected within that subject block. Thus, a single site may host several selection features. For example, the quarries at Scrabo host three selection features: the Sherwood Sandstone selected within the Triassic stratigraphy subject block, the volcanic series selected as part of the Tertiary igneous review, and finally the trace-fossil assemblage selected within Mesozoic palaeontology. The ASSI designation documentation refers to all these features, and site management activities must also address the needs of each aspect of interest.

The monitoring and management of designated sites will ensure the maintenance of scientific interest, and future review and updates of the designated site network are also essential to keep pace with new site discoveries and the development of the science itself.

Areas of Special Scientific Interest – statutory protection for ESCR sites

Designation of the ASSI series is one of the main objectives of the ESCR. The process involved is summarized below:

Table 2 ESCR progress (as at July 2001).

Theme	Features Selected	Features Designated
Stratigraphy	97	27
Igneous petrology	84	26
Palaeontology	14	4
Structural/metamorphic	24	1
Mineralogy	12	2
Pleistocene	53	6
Recent	66	45
TOTAL	350	111

- Compile information on the significant features of interest for all relevant sites, including boundary definition.
- Prioritize the order of site designation, with vulnerable or threatened localities taking precedence.
- Liaise with landowners, occupiers and other relevant parties.
- Produce descriptive citation and schedule of potentially damaging operations (the Notifiable Operations list).
- Consult with Council for Nature Conservation and the Countryside (EHS's statutory advisory body).
- Set conservation objectives defining "limits of change".
- Site declaration and confirmation.
- Site monitoring and management.
- Site condition data compiled and reported on at UK level.

The overall objective of designation is to ensure the long-term conservation of Earth science features while recognizing the legitimate rights of landowners and occupiers. However, such parties must use sites within any limits prescribed by the designation.

An example of part of a schedule is given below. The Notifiable Operations (NO) are varied, depending on the feature of interest and the degree of restraint on activities that EHS feels is necessary for long-term site protection:

- Alteration of natural or manmade features, the clearance of boulders or stones and grading of rockfaces.
- Operations or activities which would affect wetlands (including rivers, streams and open water), e.g.
  - change in the methods or frequency of routine drainage maintenance
  - modification to the structure of any watercourse
  - lowering of the water table, permanently or temporarily.
- Any change of activity or operation that involves the damage or disturbance by any means of the surface and subsurface of the land, including ploughing, rotovating, harrowing, reclamation and extraction of minerals, including rock, sand, gravel and peat.
- Sampling of rocks, minerals, fossils or any other material forming a part of the site, undertaken in a manner likely to damage the scientific interest.

The NO list does not deal with developments that fall within the sphere of planning legislation. Any such proposals are dealt with by the Planning Service, with EHS being fully consulted.

Areas of Special Scientific Interest – implications for landowners and site users

The Notifiable Operations list is the main means of safeguarding the ASSI. It places a requirement on landowners to protect the features of the site by ensuring that they consult with EHS before carrying out any of these activities, but it also has implications for third parties using such sites. A process exists by which landowners wishing to carry out a notifiable operation can apply to EHS. The likely impact of the activity on the site's scientific interest is determined and the landowner is either given consent (i.e. the operation can take place) or offered a management agreement to compensate the landowner for losses incurred by not undertaking the development.

The designation process, and the obligation that it places on landowners and occupiers to protect the site from effectively damaging activities, can have the effect of making them cautious about visitors to their land. The present legislation does not permit EHS to bring a prosecution against third parties for damage to an ASSI, perhaps sending out the signal that landowners may ultimately be prosecuted for damage caused by visitors, whether invited or not. Planned amendments to current legislation would permit third-party prosecutions and may make landowners less concerned about the activities of the public on their land.

However, EHS generally takes the view that ASSI management should not rely on threat of prosecutions but rather on discussion to resolve problems on a designated site. Further changes to the legislation will probably address access issues as they affect EHS staff. They may also make it a requirement for landowners to carry out positive site management in return for compensation payments, rather than, as is presently the case, being paid for merely agreeing not to carry out damaging activities.

Designation of Earth science localities will bring benefits by appropriately managing the potential impact of those activities that are outside the control of the planning system. It also means that EHS will be consulted for relevant development proposals within or adjoining the ASSI (e.g. new housing, quarrying, etc.). Additional gains include the opportunity to address historical site issues (e.g. *ad hoc* waste disposal, access, etc.).

Designation does not miraculously solve all site-management problems. Ultimately, protection is dependent on site owners' and users' goodwill and commitment, but designation does offer an opportunity whereby all interested parties can discuss the options for sustainable site usage.

Interim conservation status of ESCR sites

Given the relatively rapid completion of the site selection process, many sites await designation as ASSIs. In the absence of statutory protection, these important sites are vulnerable to damage, mainly through changes in site use. EHS, in collaboration with the Department of the Environment's Planning Service, has embarked upon a process of identifying sites of nature-conservation interest within regional-planning strategy documents, the Area Plans. Such plans are prepared for local authority areas by the Planning Service (the planning process in Northern Ireland is directed centrally, with a consultative role resting with local authorities) to define development and amenity zones for the life of the plan – usually some 5–10 years. Having ESCR sites included as Sites of Local Nature Conservation Importance or as Areas of Constraint on Mineral Development (the latter generally applicable to sand and gravel complexes), ensures that EHS are consulted about any proposed developments which could affect such a site. Of course, this is a means of protecting localities from activities for which planning permission is required, but it will not be a consideration for those activities that do not (e.g. afforestation or agriculture). A review of site condition is currently being undertaken of non-statutory ESCR sites, to determine if there has been any significant deterioration in site quality since they were first selected and reported on, which can, at present, be anything up to nine years.

Beyond the site-selection process

Several activities follow on from the selection and designation programmes. The Northern Ireland Earth heritage localities form an element of the United Kingdom's protected suite of sites. As such, site condition of all ASSIs must be determined on a continuing basis, to form part of the overall statement of the "health" of the UK's designated conservation sites (SSSI and ASSI). In advance of this, common standards for monitoring have been determined, to ensure broadly similar condition criteria are used in assessing all sites.

The objective of site designation is ultimately to achieve a series of site networks essential to demonstrate the geological and geomorphological development of Northern Ireland. To ensure that sites are functional in the future (gauged in part by their ability to be used as a geological resource), site-conservation objectives need to be determined. These summary plans contain statements on a site's present condition, existing and likely future threats to its integrity, management responses required in the short term, together with thresholds determined to initiate action should a site's condition fall below minimum required standards. Such thresholds may relate to the degree of vegetation encroachment or scree development obscuring an outcrop. A statement of site objectives is essential as a guide to future management and condition assessment. In addition, the plans lay down monitoring requirements and methodologies. Monitoring approaches include repeat fixed-point photography, which is appropriate for outcrop, smaller fossil, geomorphological and some low-energy geomorphological sites. Repeated vertical aerial photography is a particularly suitable technique for dynamic process sites such as coastal process systems, together with larger fossil geomorphological complexes such as deglacial landform assemblages.

Some key tasks have yet to be fully implemented. These include:

- *Site enhancement* At the designation stage, the current status of each site was assessed with a view to determining any necessary site improvements. Factors noted included ease of access to the site and to features of interest within the site. A systematic programme to implement these improvements will be required.
- *Landowner information* It was recognized at an early stage in the designation process that many landowners were not aware of the importance of Earth science localities on their land. Many such owners stated that they would be interested in receiving uncomplicated descriptions and explanations of them. Information sheets are now routinely produced

for all proposed ASSI to provide a layman's guide to the site interest. Such sheets have yet to be produced for many sites already designated.

- *Informing the geological community* It is essential that the geological community, as potential or actual site users, is informed of the implications of site designations. On occasion, such designation has been interpreted by some amateurs and professionals as meaning that these localities can no longer be used. This is not the case and such an outcome would be viewed by EHS as devaluing the site. The key consideration here is to ensure that the best sustainable use is made of a site. Although constraints are generally minimal, they are necessary: it is imperative that such users are made aware of the conservation status of these sites and told of any restrictions or limitations on site usage. Landowners will undoubtedly prove to be the main means of communicating this information to visiting geologists. The latter should also be encouraged to pass any comments relating to site condition or finds to EHS.

A website has been developed to promote the ESCR and disseminate its findings. This offers both map and keyword-based access routes into the ESCR site database. The advantages of this means of information distribution fall within those generally identified for the Web, namely ease of updating data, ease of access via keywords and links from existing Earth science websites, and the fact that it is a relatively inexpensive means of distributing the ESCR dataset. The website complements EHS's Earth science publication series, which uses the more traditional medium of print to make aspects of the ESCR programme publicly available. The website can be accessed through the Ulster Museum Habitas online site:

http://www.ulstermuseum.org.uk/habitas/

For a more complete description of the full process of the earth Science Conservation Review see Enlander (2001).

References

Ellis, N.V., D. Q. Bowen, S. Campbell, J. L. Knill, A. P. McKirdy, C. D. Prosser, M. A. Vincent, R. C. L. Wilson (eds) 1996. *An introduction to the Geological Conservation Review.* GCR Series 1, Joint Nature Conservation Committee, Peterborough.

Enlander, I. J. 2001. The Earth Science Conservation Review: conserving the earth heritage resources of Northern Ireland. *Irish Journal of Earth Sciences* **19**, 103–112.

Malvern International Task Force 1995. *Earth heritage conservation.* Joint Nature Conservation Committee, Peterborough.

Nature Conservancy Council 1990. *Earth science conservation in Great Britain – a strategy.* Nature Conservancy Council, Peterborough.

Rights of the memory of the Earth 1991: http://exodus.open.ac.uk/others/malvern/article_1.html

Further reading

References

Emeleus, C. H. & J. Preston 1969. *Field excursion guide: The Tertiary volcanic rocks of Ireland*. International Association of Volcanology and Chemistry of the Earth's Interior (published by Queen's University, Belfast).

Griffith, A. E. & H. E. Wilson 1982. *Geology of the country around Carrickfergus and Bangor* [memoir]. Belfast: GSNI.

Harland, W. B. 1969. Tectonic evolution of the North Atlantic region. In *North Atlantic: geology and continental drift*. Memoir 12, American Association of Petroleum Geologists, 817.

Hammond, F. 1991. *Industrial heritage – Antrim coast and glens*. Belfast: HMSO.

Legg, I. C., T. P. Johnston, W. I. Mitchell, R. A. Smith 1998. *Geology of the country around Derrygonnelly and Marble Arch* [memoir]. Belfast: GSNI.

Long, C. B. & B. J. McConnell 1997. *Geology of north Donegal* [memoir]. Dublin: GSI.

Long, C. B. & B. J. McConnell 1999. *Geology of south Donegal* [memoir]. Dublin: GSI.

Lyle, P. 1996. *A geological excursion guide to the Causeway Coast*. Belfast: Environment and Heritage Service.

Lyle, P. & J. Preston 1993. The geochemistry and volcanology of the Tertiary basalts of the Giant's Causeway area, Northern Ireland. *Journal of the Geological Society* **150**, 109–120.

Macdermott, C. V., C. B. Long, S. J. Harney 1996. *Geology of Sligo–Leitrim* [memoir]. Dublin: GSI.

Sharpe, E. N. 1970. An occurrence of pillow lavas in the Ordovician of County Down. *Irish Naturalists Journal* **16**, 299–301.

Smith, R. A., T. P. Johnston, I. C. Legg 1991. *Geology of the country around Newtownards* [memoir]. Belfast: GSNI.

Wilson, H. E. 1972. *The regional geology of Northern Ireland*. Belfast: HMSO.

Wilson, H. E. & P. I. Manning 1978. *Geology of the Causeway Coast*, vol. 1 [memoir]. Belfast: GSNI.

Wilson, H. E. & J. A. Robbie 1966. *Geology of the country around Ballycastle* [memoir]. Belfast: GSNI.

Woodcock, N. H. & R. A. Strachan (eds) 2000. *Geological history of Britain and Ireland*. Oxford: Blackwell Science.

Wyllie, P. J. 1976 *The way the Earth works: an introduction to the new global geology and its revolutionary development*. New York: John Wiley.

The geological survey organizations

Geological maps, memoirs and other services are provided in Northern Ireland by the Geological Survey of Northern Ireland (GSNI) and in the Republic of Ireland by the Geological Survey of Ireland (GSI).

Geological Survey of Northern Ireland
20 College Gardens, Belfast BT9 6BS
Telephone +44 (0)28 90 666595
www.bgs.ac.uk/gsni

Geological Survey of Ireland
Beggars Bush, Haddington Road, Dublin 4
Tel. +353 1 6707444
www.gsi.ie

Maps: GSNI

1:250 000 – *Geological map of Northern Ireland*

1:50 000 – *Ballycastle, Ballymena, Causeway Coast, Coleraine, Derrygonnelly and Marble Arch, Enniskillen, Mourne Mountains, Draperstown*

1:63 360 – *Belfast, Carrickfergus* [includes north Down]

Maps: GSI
1:100 000 series
North Donegal, sheet 1 and part of sheet 2
South Donegal, sheet 3 and part of sheet 4
Sligo–Leitrim, sheet 7

Glossary

absolute age The dating of geological events in years.

adit A horizontal or near-horizontal passage by which a mine is entered.

agglomerate A rock composed of coarse fragmental debris produced by explosive volcanic activity.

ammonite One of a large group of extinct sea animals related to the modern *Nautilus*.

amygdale A gas cavity or vesicle in volcanic rocks which has become filled with secondary minerals, for example zeolites.

amygdaloidal lava Lava with many small gas cavities, usually filled with minerals such as zeolites.

andesite A volcanic rock intermediate between basalt and rhyolite.

aquifer A water-bearing layer of rock.

ash Material blown out of an explosive volcano; when consolidated into rock it is called tuff.

asthenosphere The layer within the Earth below the lithosphere, a relatively weak zone characterized by low seismic velocities, and probably partially molten.

basalt Dark fine-grain basic igneous rock.

batholith A large mass of intrusive igneous rock formed deep within the Earth.

bauxite A lateritic rock rich in hydrated aluminium oxides; used as an ore of aluminium and a source of aluminium compounds, particularly aluminium sulphate.

belemnite An extinct type of cephalopod or squid-like creature known from cigar-shape fossils found in the Cretaceous Chalk.

bentonite A clay formed from the decomposition of volcanic ash.

boulder clay Ground moraine produced by the grinding action at the base of a moving ice sheet; consists of a stiff clay containing stones and boulders of all sizes.

breccia A rock composed of angular fragments embedded in a matrix, produced by the breaking up of a consolidated rock by some natural agency such as faulting, collapse or volcanic activity.

Bronze Age The archaeological period when bronze tools, weapons and ornaments were produced, 3000–2500 BC in the north of Ireland.

Cainozoic The last of the four eras into which geological time is divided, as recorded by the stratigraphical rocks of the Earth's crust.

Caledonian orogeny Mountain-building period caused by the closure of the Iapetus Ocean in Silurian and early Devonian times.

Carboniferous Period of the stratigraphical column after the Devonian and before the Permian, approximately 345–280 million years ago.

chert A form of extremely fine-grain silica, commonly distributed through limestone; similar in composition to flint.

clastic Consisting of fragments of rock that have been moved from their place of origin.

clint Ridges or blocks of limestone, separated by solution joints or fissures; see also grykes.

coccolith A type of marine algae with a calcareous shell.

colonnade The zone of regular columns at the base of a multi-tier columnar lava of Causeway Tholeiite type; from the architectural term referring to a range of columns placed at regular intervals and supporting the entablature.

columnar jointing The division of an igneous rock into columns by cracks produced as a result of thermal contraction or shrinkage on cooling.
conchoidal fracture A type of rock or mineral fracture that produces a smoothly curved, shell-like surface; a characteristic of flint, for example.
cone sheet Funnel-shape dyke, generally concentric, around an igneous intrusion.
conformable succession A sequence of sedimentary rocks that indicates continuous deposition, with no erosion, over a geologically long period.
conglomerate A sedimentary rock composed of rounded fragments of gravel or pebble size.
constructive plate margin In plate tectonics, a plate margin where plates are diverging or moving apart and the crust is under tension with new crust being created.
contact metamorphism The changes in the mineral composition and texture of a rock produced by contact with intrusive or extrusive igneous rocks.
continental drift The horizontal displacement or rotation of continents relative to one another, an essential process in plate tectonics.
convection A mechanism of heat transfer in a flowing material in which hot material from the bottom rises because of its lower density, while cool surface material sinks.
convection cell A single closed flow circuit of rising warm material and sinking cold material.
core The central part of the Earth below a depth of 2900 km; thought to consist of iron and nickel and to be molten on the outside with a solid inner section.
Cretaceous Period of the stratigraphical column after the Jurassic and before the Tertiary, from approximately 145 million years to 65 million years ago.
crinoids A group of marine fossils related to the echinoderms, consisting of a cup-like structure with radiating arms raised above the sea floor by an elongated jointed stem.
crust The outermost layer of the lithosphere; the continental crust is mostly granite in composition and the oceanic crust is composed mainly of basaltic rocks.

Dalradian Supergroup A belt of metamorphic rocks that stretches from Connemara, north to Donegal, Tyrone, northeast Antrim and into the Highlands of Scotland; they range in age from late Precambrian (Neoproterozoic) to early Ordovician.
declination The angle between geomagnetic north and geographical north at any point on the Earth's surface.
destructive plate margin In plate tectonics, a plate margin where plates are converging or moving together and the crust is under compression and therefore being shortened.
Devonian The period of the stratigraphical column after the Silurian and before the Carboniferous, from approximately 395 million years to 345 million years ago.
diorite A plutonic rock with composition intermediate between gabbro and granite, the intrusive equivalent of andesite.
dolerite Medium-grain dark basic rock, of the same composition as basalt but with coarser crystals.
drift A collective term for all the rock, sand and clay that has been transported and deposited by a glacier, or by water derived from ice melting.
drift map A geological map showing the distribution of the drift or superficial deposits.
drumlin A streamlined hill composed of glacial till with its long axis parallel to the direction of ice flow.
drusy Containing many druses, which are small cavities lined with crystals.
dyke Vertical or near-vertical wall-like sheet of intrusive igneous rock that has discordant contacts with the surrounding rock.

effusive A form of volcanic eruption where the lava flows consist of a low-viscosity magma such as basalt.
entablature The zone of narrow, often curved, columns above the colonnade columns in a multi-tiered columnar lava of Causeway Tholeiite type. From

the architectural term referring to that part of the building which rests on the colonnade and includes the frieze.

eon The largest division of geological time, covering several eras, for example the Phanerozoic, 600 million years ago to the present day.

era A division of geological time including several periods.

esker Sand and gravel in the form of a narrow ridge deposited by meltwater channels from a glacier or icecap.

extrusive igneous rock Material resulting from volcanic activity at the Earth's surface, generally solidified lava.

fault A fracture or zone of fractures along which differential movement of the wallrocks has taken place.

fault breccia A rock composed of coarse angular fragments resulting from crushing and grinding along faults.

faunal succession The sequence of life forms as recorded by the fossil remains in a stratigraphical succession.

fissure eruption Volcanic eruption from a linear fissure or crack in the Earth's surface, usually with the emission of basic lava.

flint A form of microscopically crystalline silica occurring as nodules in limestone; similar in composition to chert.

foliation Any planar set of minerals or banding of mineral concentrations found in a metamorphic rock.

formation In stratigraphy, the primary unit consisting of a succession of rocks useful for mapping or description, e.g. the Lower Basalt Formation.

fracture The manner of breaking and the appearance of a mineral when broken, e.g. conchoidal fracture; or breaks in rocks caused by to faulting or folding.

gabbro A dark coarse-grain intrusive igneous rock, the intrusive equivalent of basalt.

geomagnetic field The magnetic field of the Earth.

geomorphology The study of landforms.

glauconite Dark-green mineral – a hydrous silicate of iron and potassium – occurring locally in the Cretaceous rocks and taken to indicate a slow rate of deposition.

gneiss A coarse-grain regional metamorphic rock that shows compositional banding of minerals.

Gondwanaland Theoretical supercontinent comprising approximately the present continents of the Southern Hemisphere, considered to have fragmented and drifted apart from Permian times onwards.

graptolite Extinct planktonic colonial animals, used in understanding the stratigraphy of the Ordovician and Silurian in Britain and Ireland.

granite A pale coarse-texture igneous rock, essentially made up of quartz, feldspar and mica, the intrusive equivalent of rhyolite.

greywacke A poorly sorted sandstone containing abundant feldspar and rock fragments, in a clay-rich matrix.

group In stratigraphy, a unit consisting of two or more formations.

gryke Solution-widened fissures in limestone areas; see also clint.

Holocene In stratigraphy, the younger or recent part of the Quaternary period.

hornfels Tough fine-grain rock produced as a result of the thermal metamorphism of sediments by the heat from an igneous intrusion.

hyaloclastite Fragments of glassy material formed by the chilling or quenching of basaltic magma in water during subaqueous eruptions.

Iapetus Ocean Earlier version of the Atlantic Ocean, which existed between 600 million years and 400 million years ago.

Iapetus suture The line marking the collision of the continental masses of Laurentia and Avalonia as the Iapetus Ocean closed.

igneous rock Rock formed by the crystallization of molten material.

inclination The angle between a line in the Earth's magnetic field and the horizontal.

intrusive rock A body of molten igneous rock that has intruded or invaded

older rocks beneath the surface, and subsequently solidified.
island arc A linear chain of volcanic islands formed at a convergent plate boundary.

Jurassic Period of the stratigraphical column after the Triassic and before the Cretaceous, from approximately 210 million years to 145 million years ago.

lamprophyre Dyke rocks intermediate in composition between basalt and granite, containing biotite, pyroxene and hornblende.
laterite Weathering product, particularly of igneous rocks, under wet tropical conditions. The formation of clay minerals such as kaolin is followed by removal in solution of silica and other elements, leaving a characteristically red mixture of iron and aluminium oxides.
Laurentia Supercontinent comprising approximately the present continents of the Northern Hemisphere, considered to have fragmented and drifted apart from Permian times onwards.
lava Extruded igneous rock, commonly in the form of flat sheets (lava flows).
Lias In stratigraphy, part of the lower Jurassic period, characterized in north-east Ireland by grey clays and mudstones.
lignite A low-rank coal, dark brown in colour, formed by the partial decomposition of vegetable matter under anaerobic conditions.
limestone A sedimentary rock mostly composed of the calcium-carbonate mineral, calcite.
limestone pavement Bare limestone surface fretted by solution.
lithosphere The outer rigid shell of the Earth above the asthenosphere and comprising the crust and the upper brittle part of the mantle.

magma Molten rock material that forms igneous rocks on cooling. Magma that reaches the surface is extrusive and forms lava; magma that solidifies below the surface forms intrusive igneous rocks.
magnetic North Pole The point where the Earth's surface intersects the axis or line of the dipole that approximates the Earth's magnetic field and where this field dips vertically downwards.
magnetite Magnetic iron ore.
mantle The portion of the Earth between the crust and core.
mantle convection The system of convection cells within the mantle thought to be the principal driving force for plate tectonics.
mantle plumes Rising jets of hot, partially molten material from within the mantle and thought to be responsible for volcanism occurring away from plate boundaries such as in Hawaii.
marble The metamorphic equivalent of limestone.
megalith A Neolithic monument constructed from large stones.
member In stratigraphy a division of a formation, generally of distinct character or only local extent.
Mesolithic Literally, the "middle stone age"; in Ireland from about 8000–5000 years ago.
Mesozoic The era of geological time between the Palaeozoic and the Cainozoic.
metamorphism Change in the character of a rock attributable to heat or pressure, or both, with the development of new minerals and structures.
microlith Small tools often made from flint or chert, characteristic of the Mesolithic in Ireland.
migmatite A rock, with both igneous and metamorphic characteristics, that shows laminar flow structures.
mid-ocean ridge A major elevated linear feature of the sea floor; a characteristic type of constructive plate boundary where plates are diverging.
moraine A glacial deposit of till left at the margins of an ice sheet or glacier
mudstone A sedimentary rock made of clay-size particles; structureless and unlaminated.
mylonite A very fine fault breccia commonly associated with major thrust faults and produced by shearing during fault movement.

Neogene In stratigraphy the period between the Palaeogene and the

Quaternary, formerly the upper part of the Tertiary.

Neolithic Literally, the "new stone age"; in Ireland from 3000 BC to 1750 BC.

obduction The process occurring during plate collision whereby a piece of the subducted plate is broken off and thrust upwards onto the overriding continental plate, to produce an ophiolite complex.

obsidian Dark volcanic glass formed when lava of rhyolite composition cools quickly without forming crystals.

Old Red Sandstone Continental sandstones of the Devonian period.

ophiolite complex An assemblage of mainly basaltic igneous rocks and deep-sea sediments, considered to represent slices of oceanic crust and upper mantle, that have been thrust or obducted upwards onto continental crust.

Ordovician Period of the stratigraphical column after the Cambrian and before the Silurian, approximately 510–440 million years ago.

Palaeogene In stratigraphy, the period between the Cretaceous and the Neogene, formerly the lower part of the Tertiary period.

Palaeozoic In stratigraphy, the era of ancient life, a subdivision of the Phanerozoic eon.

Pangaea A supercontinent that was formed of all the present continents and which existed about 200 million years ago.

pegmatite An igneous rock with extremely large grains, often found as veins or dykes in plutonic rocks of finer grain size.

pelite Often applied to fine-grain metamorphic rocks derived from sediments rich in clay minerals, such as shales and mudstones.

period In stratigraphy, a major worldwide unit of geological time.

Permian Period of the stratigraphical column after the Carboniferous and before the Triassic, from approximately 280 million years ago to 245 million years ago.

Phanerozoic In stratigraphy, the eon of evident life, approximately from Cambrian to Recent.

phenocryst In igneous rocks, a relatively large crystal surrounded by a finer-grain matrix.

phreatic tube Passage formed in limestone areas when all the spaces in the rock are completely filled with water and erosion takes place around the whole circumference to produce a smooth round outline.

pillow lavas A type of lava formed under water in which many small pillow-shape tongues break through the chilled surface and quickly solidify.

plankton Floating organisms in oceans or lakes.

plate One of the dozen or more segments of the lithosphere that move independently over the surface of the Earth, meeting at convergent or destructive margins and separating at divergent or constructive margins.

plate tectonics, theory of The theory and study of the formation, movement and interaction of lithospheric plates on the Earth's surface.

Pleistocene In stratigraphy, the older subdivision of the Quaternary period.

pluton A large igneous intrusion formed at depth in the crust.

porcellanite Rock composed of thermally metamorphosed clay-rich rocks.

porphyritic In igneous rocks, a texture in which larger crystals (phenocrysts) are set in a finer-grain matrix.

Precambrian In stratigraphy, all rocks formed before the Cambrian period, i.e. older than approximately 500 million years.

Proterozoic In stratigraphy, the eon of first life forms.

psammite Often applied to medium-grain metamorphic rocks having the composition of sandstones.

quartz A common mineral in granitic rocks, because of its hardness and durability it is the main component of sands and sandstones.

quartzite A hard rock formed by the metamorphism of quartz-rich sandstone.

Quaternary Period of the stratigraphical column after the Neogene, from approximately 2 million years ago to Recent.

radioactive decay The breakdown of certain heavy elements to lighter elements by the emission of charged particles or radiation.
radiometric dating The method of obtaining ages of geological materials by measuring the relative abundances of a radioactive element and the lighter element produced from it by radioactive decay.
raised beach Coastal beaches and cliffs above the present shoreline, cut at a time after the Ice Age when sea level was higher because of the melting of the icecaps.
regional metamorphism Metamorphism occurring over a wide area and caused by deep burial or the tectonic forces of the Earth.
regolith Any solid material lying on top of the bedrock and including material weathered from it.
relative age The dating of geological events according to the system of successive eras, periods and epochs used in geology and palaeontology.
rhyolite Fine-grain acid lava; the extrusive equivalent of granite.
rifting Faulting caused by divergence or tension in the crust.

sandstone Sedimentary rock composed of consolidated and cemented sand grains.
schist A metamorphosed rock, characterized by the development of secondary mica, which gives it a lustrous appearance.
seafloor spreading The mechanism by which new seafloor crust is created at mid-ocean ridges in constructive plate margins as adjacent plates are moved apart to make room.
sedimentary rock A rock formed by the accumulation and consolidation of mineral fragments transported by wind, water or ice to the deposition site, or formed by chemical action at the deposition site.
shale A fine-grain sedimentary rock of clay or silt grade, with pronounced layering or lamination.
sill Flat-lying mass of igneous rock intruded conformably with the layering of the rocks into which it has been injected.
siltstone A fine-grain sedimentary rock of silt-grade particles, not conspicuously bedded.
Silurian Period of the stratigraphical column after the Ordovician and before the Devonian, about 440–410 million years ago.
sinkholes A small steep depression caused by the dissolution and collapse of subterranean caves in limestone areas.
slate A fine-grain metamorphic rock derived from mudstone or shale and possessing a well developed foliation.
sole structures Features formed on the base or lower surface of a sedimentary bed, often in environments were turbidity currents were common, and which can be used to determine the way-up of a sediment.
solid map A geological map showing the distribution of the solid or bedrock geology.
spheroidal weathering The formation of spherical boulders by chemical weathering along joints; commonly found in weathered basalts.
stack A high rock off the coast, detached by marine erosion from the main cliff.
stratigraphy The science of description, correlation and classification in sedimentary rocks, including the interpretation of the depositional environment of those rocks.
stratigraphical column The division of geological history into smaller units such as eras and periods through stratigraphy and the study of fossils.
subduction In plate tectonics, the process whereby the oceanic plate is forced down into the mantle under the continental plate at a destructive or converging plate margin.
supergroup In stratigraphy, several related groups, for example the Dalradian.
superposition, law of The principle that, in a sequence of undeformed sediments, a bed that overlies another bed is always the younger.
swarm In igneous petrology, the term given to a group of dykes of similar orientation and composition, and the same age of intrusion.

tectonics The study of the broad structural features of the Earth, and their causes.

terrane A fault-bounded geological entity characterized by a distinctive stratigraphical or structural history, or both, which differs markedly from the surrounding areas.

Tertiary Period of the stratigraphical column after the Cretaceous and before the Quaternary, approximately 65–2 million years ago.

thermal metamorphism See contact metamorphism.

thrust fault A fault caused by compression in which the fault plane is at a low angle of inclination to the horizontal.

Triassic Period of the stratigraphical column after the Permian and before the Jurassic, approximately 250–210 million years ago.

tuff A rock formed by compaction of fragmental volcanic debris such as ash.

turbidite Sedimentary deposit formed by a turbidity current.

turbidity current A mass of mixed water and sediment that flows down the continental slope of an ocean because it is denser than the surrounding water.

unconformable Sedimentary beds not succeeding the underlying strata in immediate order of age and not lying parallel to them.

unconformity A surface of erosion or non-deposition separating younger rocks from older.

uniformitarianism, principle of The concept that the processes that shaped the Earth throughout geological time are the same as those observable today.

vent A volcanic pipe or opening filled with fragmental material.

vesicle A small gas cavity, usually spherical, in a volcanic rock.

xenolith A piece of the surrounding country rock found included in an igneous intrusion.

zeolites A group of minerals composed of the hydrated silicates of calcium and aluminium, sometimes with sodium and potassium; commonly found in vesicles in the Antrim basalts.

Index